NORTH CAROLINA
STATE BOARD OF COMMUNITY COLLEGES
LIBRARIES
ASHEVILLE-BUNCOMBE TECHNICAL COMMUNITY COLLEGE

FUNDAMENTALS OF
DIRECT CURRENT

DISCARDED

JUN 2 3 2025

DEDICATED TO TERRY
Who kept me going.

FUNDAMENTALS OF
DIRECT CURRENT

ROBERT E. ARMSTRONG

TAB BOOKS Inc.
Blue Ridge Summit, PA 17214

FIRST EDITION

FIRST PRINTING

Copyright © 1986 by TAB BOOKS Inc.

Printed in the United States of America

Reproduction or publication of the content in any manner, without express permission of the publisher, is prohibited. No liability is assumed with respect to the use of the information herein.

Library of Congress Cataloging in Publication Data

Armstrong, Robert E.
   Fundamentals of direct current.

   Includes index.
   1. Electric engineering. 2. Electric currents,
Direct. I. Title.
TK1111.A76 1986   621.319′12   85-27632
ISBN 0-8306-0870-2
ISBN 0-8306-1870-8 (pbk.)

Contents

Introduction — vii

1 Electron Physics — 1
Matter—Molecules—Atoms—Electrons—Conductors—Insulators—Static Electricity—Law of Attraction and Repulsion—Electrostatic Field—Work—Energy—Power—Efficiency—Quiz

2 Electrical Properties and Basic Measurement Units — 17
Polarity—Emf and the Volt—Sources of Emf—Current Flow and the Ampere—Resistance and the Ohm—The Simple Circuit—Basic Schematic Symbols—Quiz

3 Dc Circuit Analysis — 31
Mathematics for Electricity—Prefixes—Ohm's Law—Electrical Circuits—Series Circuits—Parallel Circuits—Series-Parallel Circuits—Circuit Analysis—Quiz—Answers to Exercises

4 Magnetism — 59
Retentivity—Magnetic Poles—The Magnetic Earth—Theory of Magnetism—Law of Attraction and Repulsion—Magnetic Force and Its Measurement—Magnetic Field—Shapes and Uses of Magnets—Care of Magnets—Electromagnetism and Magnetic Circuits—Electric and Magnetic Circuit Properties Compared—Quiz

5 Batteries — 83
The Primary Cell—Battery Combinations—Secondary Cells—Quiz

6 Resistors 107
Types of Resistors—Resistor Values—The Resistor Color Code—Temperature Coefficient—Quiz

7 Dc Measuring Devices 117
The Ammeter—The Voltmeter—The Ohmmeter—The Multimeter—Quiz

8 Capacitance 135
The Property of Capacitance—Theory of Capacitance—Capacitive Time Constant—Factors Affecting Capacitance—Calculating Total Capacitance—Types of Capacitors—Special Types of Capacitors—Color Code—Quiz

9 Inductance 159
Self-Inductance—Factors Affecting Inductance—Effects in Dc—Inductive Time Constant—Mutual Inductance—Calculating Total Inductance—Inductor Construction—Uses for Inductors—Quiz

Appendices 181
A. Answers to Quizzes 181
B. Dc Formulas 185
C. Electrical Device Reference Designations 187
D. Abbreviations for Electrical Devices 191
E. Wire Gauge Sizes and Current Capacity 197
F. Symbols for Electrical-Electronic Devices 199

Glossary 217

Index 239

Introduction

Electronics can be defined as that portion of the field of electricity devoted exclusively to the behavior and movement of electrons through all types of devices such as electron tubes, semiconductors, inductors, and capacitors. Some major branches of electronics are communications, navigation, radar, sonar, and television. Understanding the fundamentals of direct current electricity is a definite prerequisite to any serious understanding of electronics. This book provides electrical novices with a strong foundation for more advanced work in electronics.

While teaching a course in basic electricity and electronics at the U. S. Navy Radioman School many years ago, I saw that the most important principle for the student to grasp is the simple, logical understanding of what electricity actually is. Once this principle is grasped, the study of electricity and electronics is much easier and more interesting. To provide a well-balanced text and utilize your fullest learning potential, we have minimized the use of mathematics and made every effort to maintain a close relationship between theory and practice. To help you master the material, we have included sample problems and quizzes at important points in each chapter, which illustrate the application of basic principles to problem solving.

So sharpen your pencil and break out the scratch pad. Let's explore the fascinating world of the electron.

Chapter 1

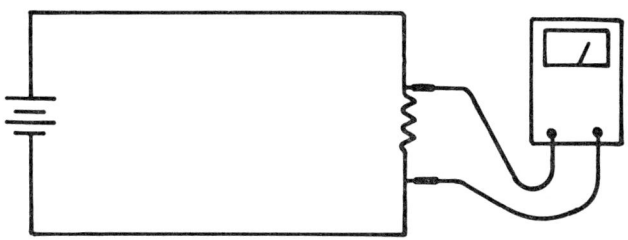

Electron Physics

PHYSICS IS ONE OF THE MANY SCIENCES THAT HAS CONTRIButed heavily to our understanding of the nature and knowledge of electricity and electronics. Many fundamental principles of physics are directly related to the study of electricity and electronics. These principles have existed since the beginning of time, but they have been discovered only fairly recently. These discoveries by early scientists led to observations of some basic behavior patterns of electricity, or the electric effect, and a theory was formulated. Later, through much experimental verification of this *theory*, scientists were able to determine precise physical *laws*. Using these laws, man has been able to transform this electric effect into different forms of energy—mainly light energy, heat energy, sound energy, and mechanical energy. Modern devices such as electric motors, television sets, radios, and light bulbs are examples of how this energy is used.

Modern civilization has become dependent on these devices to maintain production output and high standards of living. Because this chapter contains the necessary ingredients to understand this electric effect, you should treat it as the *most important* chapter of the book.

MATTER

The basic structure of everything on this earth conforms to a set pattern called the *atomic structure* of matter. *Matter* is anything that has mass and occupies space. For example, to construct a building, you must use proper materials and apply them in the proper order. These materials then form a building. When we consider the basic structure of all matter, we can use the same concept because there is also a definite order in nature.

Matter may consist of a single element or two or more elements, and it can have various forms: it may be a gas, a liquid, or a solid. For example, in one environment water is a liquid; in another, it becomes a solid (ice); and in still another, it becomes a gas (steam).

Chemists and physicists have determined that there are 92 naturally existing elements along with 11 man-made ones. Combining different elements produces many different materials. For example, combining iron, manganese, and carbon produces steel.

The elements and their atomic weights are listed in Table 1-1.

MOLECULES

Now you need to exercise your imagination! If we could take any of these elements and break it down into microscopic and submicroscopic parts, we would finally reach the point where it couldn't be broken down any further; if it could, then it would no longer be the element we started with. When we reach this limit, we are looking at a new object, called a molecule. A *molecule* is the smallest particle to which an element can be reduced and still retain its original characteristics. Consequently, it is referred to as a building block of matter.

As we descend into the world of matter, we see that things we perceived as being solid are not. We also see that molecules differ in appearance: a paper molecule is different from an iron molecule, and a water molecule is different from a wood molecule.

ATOMS

Descending even further into matter, we see that these molecules become huge—we can even see what they are made of. They remind us of a galaxy in the universe—blazing suns with smaller planet-like particles whirling around them. We have now reached our goal!

An *atom* is the smallest particle to which an element can be reduced and still retain its original characteristics. Examining Fig. 1-1, we see that an atom consists of two distinct parts. Although this represents the simplest atom of all elements, the basic makeup of all atoms is pretty much the same. Atoms comprise a heavy nu-

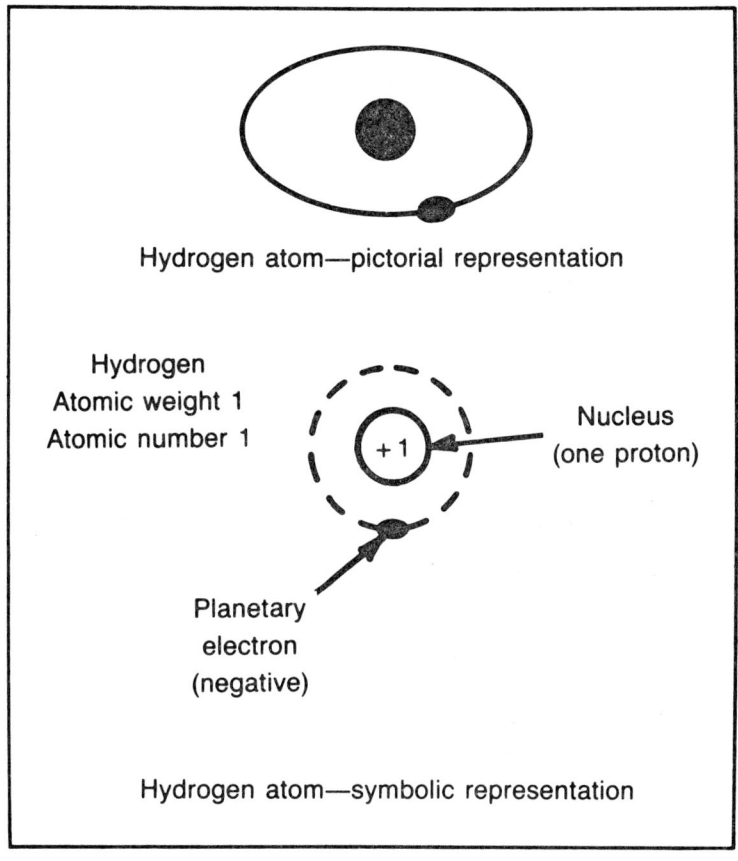

Fig. 1-1. Methods of representing atoms.

cleus in the center called a *proton*, which is a positively charged particle.

ELECTRONS

The smaller, whirling, planet-type of particle is called an *electron*, which is a negatively charged particle. We are keenly interested in the electron because we shall be analyzing them very closely by measuring, counting, predicting and controlling their behavior.

An electron is quite small and has never actually been seen by the human eye. Scientists assume they exist by the tracks they leave on photographic plates. This observation also says that, for our simple hydrogen atom in Fig. 1-1, the electron is 1850 times lighter

Table 1-1. Chemical Elements.

Element & Symbol	Atomic Number	Atomic Weight
actinium (Ac)	89	
aluminum (Al)	13	26.9815
americium (Am)	95	
antimony (Sb)	51	121.75
argon (Ar)	18	39.948
arsenic (As)	33	74.9216
astatine (At)	85	
barium (Ba)	56	137.34
berkelium (Bk)	97	
beryllium (Be)	4	9.01218
bismuth (Bi)	83	208.9806
boron (B)	5	10.81
bromine (Br)	35	79.904
cadmium (Cd)	48	112.40
calcium (Ca)	20	40.08
californium (Cf)	98	
carbon (C)	6	12.011
cerium (Ce)	58	140.12
cesium (Cs)	55	132.9055
chlorine (Cl)	17	35.453
chromium (Cr)	24	51.996
cobalt (Co)	27	58.9332
copper (Cu)	29	63.546
curium (Cm)	96	
dysprosium (Dy)	66	162.50
einsteinium (Es)	99	
erbium (Er)	68	167.26
europium (Eu)	63	151.96
fermium (Fm)	100	
fluorine (F)	9	18.9984
francium (Fr)	87	
gadolinium (Gd)	64	157.25
gallium (Ga)	31	69.72
germanium (Ge)	32	72.59
gold (Au)	79	196.9665
hafnium (Hf)	72	178.49
helium (He)	2	4.00260
holmium (Ho)	67	164.9303
hydrogen (H)	1	1.0080
indium (In)	49	114.82
iodine (I)	53	126.9045
iridium (Ir)	77	192.22
iron (Fe)	26	55.847
krypton (Kr)	36	83.80
lanthanum (La)	57	138.9055
lawrencium (Lr)	103	
lead (Pb)	82	207.2
lithium (Li)	3	6.941
lutetium (Lu)	71	174.97
magnesium (Mg)	12	24.305

Element & Symbol	Atomic Number	Atomic Weight
manganese (Mn)	25	54.9380
mendelevium (Md)	101	
mercury (Hg)	80	200.59
molybdenum (Mo)	42	95.94
neodymium (Nd)	60	144.24
neon (Ne)	10	20.179
neptunium (Np)	93	237.0482
nickel (Ni)	28	58.71
niobium (Nb)	41	92.9064
nitrogen (N)	7	14.0067
nobelium (No)	102	
osmium (Os)	76	190.2
oxygen (O)	8	15.9994
palladium (Pd)	46	106.4
phosphorous (P)	15	30.9738
platinum (Pt)	78	195.09
plutonium (Pu)	94	
polonium (Po)	84	
potassium (K)	19	39.102
praseodymium (Pr)	59	140.9077
promethium (Pm)	61	
protactinium (Pa)	91	231.0359
radium (Ra)	88	226.0254
radon (Rn)	86	
rhenium (Re)	75	186.2
rhodium (Rh)	45	102.9055
rubidium (Rb)	37	85.4678
ruthenium (Ru)	44	101.07
samarium (Sm)	62	150.4
scandium (Sc)	21	44.9559
selenium (Se)	34	78.96
silicon (Si)	14	28.086
silver (Ag)	47	107.868
sodium (Na)	11	22.9898
strontium (Sr)	38	87.62
sulfur (S)	16	32.06
tantalum (Ta)	73	180.9479
tachnetium (Tc)	43	98.9062
tellurium (Te)	52	127.60
terbium (Tb)	65	158.9254
thallium (Tl)	81	204.37
thorium (Th)	90	232.0381
thulium (Tm)	69	168.9342
tin (Sn)	50	118.69
titanium (Ti)	22	47.90
tungsten (W)	74	183.85
uranium (U)	92	238.029
vanadium (V)	23	50.9414
xenon (Xe)	54	131.30
ytterbium (Yb)	70	173.04
yttrium (Y)	39	88.9059
zinc (Zn)	30	65.37
zirconium (Zr)	40	91.22

than the proton it circles. They estimate it would take 500,000,000,000,000,000,000,000,000 electrons to make a pound of any substance!

Figure 1-2 illustrates the more complex copper atom. This atom contains more electrons than our simple hydrogen atom. Note the single electron located in the fourth (outermost) ring.

The attraction between the electrons located in the first, second, and third rings, and nucleus is sufficient to keep these electrons, called *bound* electrons, in their orbits around the nucleus. The single electron orbiting the nucleus in that fourth ring, however, is far enough away from the nucleus so that this attraction is not strong enough to keep it bound. Any type of outside force can knock it free of its orbit, and it will drift away and take up another orbit in the outermost or fourth ring of an adjacent atom either by simply filling the space vacated by that atom's single electron or by knocking that single electron out of its orbit and taking its place. This movement of *free* electrons within any material is the basis of all "electrical effect" or *electricity* as we know it.

Figure 1-3 shows this electron movement from atom to atom.

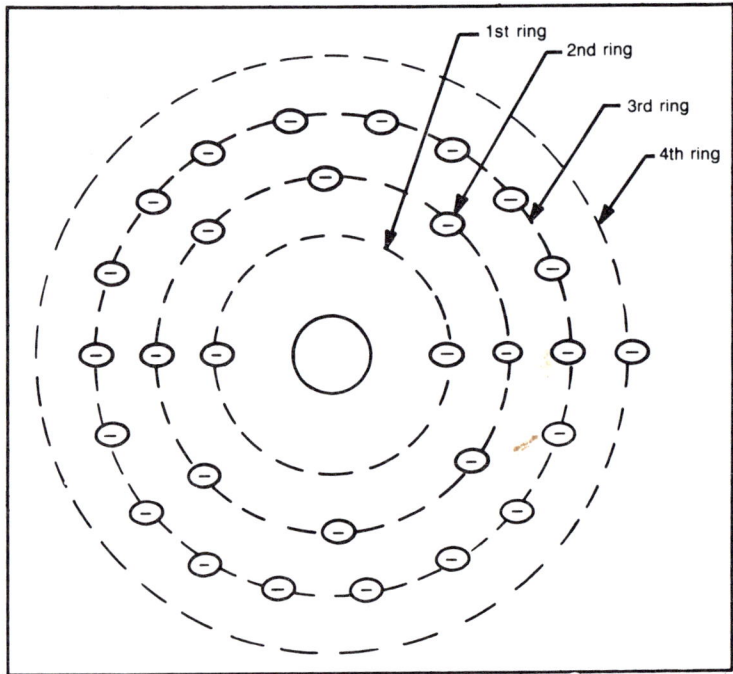

Fig. 1-2. Symbolic representation of a copper atom.

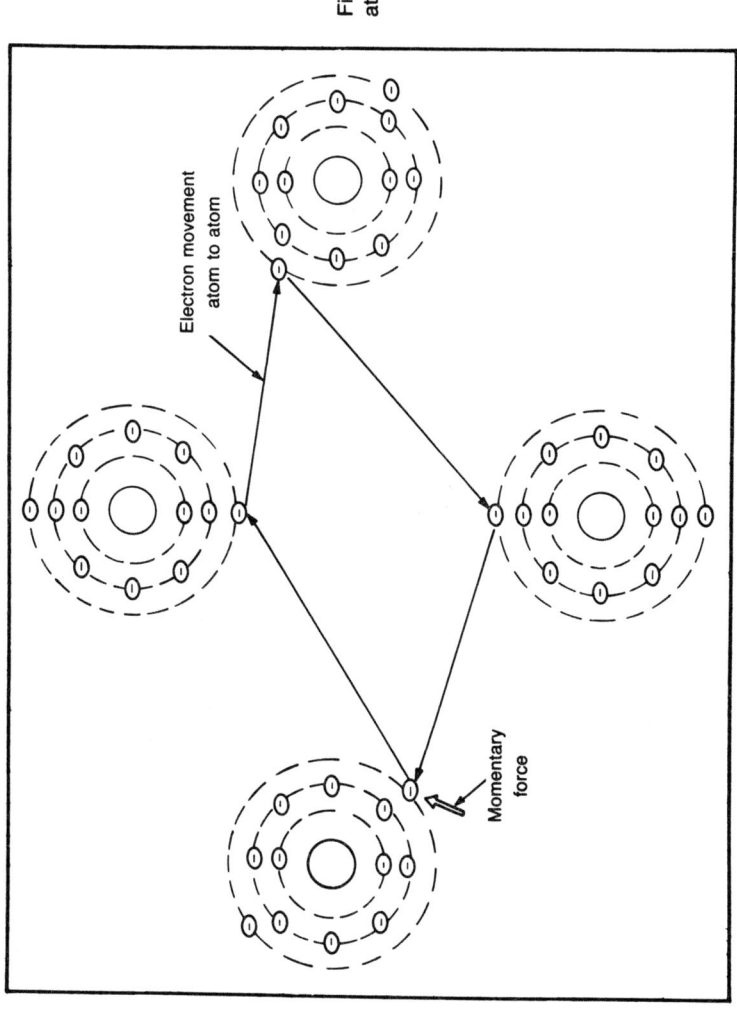

Fig. 1-3. Electron movement from atom to atom.

You must realize, however, that within any material there are trillions of these atoms bunched together, even though our illustration shows only four.

CONDUCTORS

As we study the atomic structure of matter we can readily see that each atom has its own characteristics. Some atoms contain free electrons, others contain none. Therefore, some materials have many free electrons, and other materials have none.

A material composed of atoms with free electrons is called a *conductor*. A conductor allows the free electrons to flow quite easily. All metals are conductors, some much better than others. Silver, copper, and aluminum are commonly used conductors. Silver is the best, but its cost precludes its general use in most applications. It is used in some specialized areas such as relay contacts.

Copper is the most common conductor because it has good conductivity and relatively low cost. Table 1-2 compares the conductibility of various materials.

INSULATORS

An *insulator* is a material with very few or no free electrons.

Table 1-2. Relative Conductances of Various Materials.

Material	Relative Conductance
Silver	1.08
Copper	1.00
Gold	0.725
Aluminum	0.629
Tungsten	0.312
Zinc	0.275
Brass	0.227
Platinum	0.172
Iron	0.149
Nickel	0.129
Tin	0.121
Steel	0.116
Lead	0.081
Manganin	0.0385
Mercury	0.018
Nichrome	0.0166
Carbon	0.0004

No material is a perfect insulator. Any material can be forced to permit a flow of electrons if enough external force is applied. When this happens, the insulator is said to have "broken down." Some common insulating materials are glass, porcelain, and rubber.

STATIC ELECTRICITY

Experimentation with moving electrons was actually carried on many centuries ago. Around 600 B.C. the Greek philosopher Thales observed that strange things happened when he rubbed a piece of amber with a piece of fur. After rubbing the amber with the fur, the amber attracted itself to many types of lightweight objects, such as straw, feathers, pieces of lint, and particles of dust. Thales did not quite understand what he was dealing with. He thought it was a phenomena dealing strictly with the properties of amber and fur.

It was not until the eighteenth century that the modern theory of static electricity evolved, and it was discovered that this same phenomena observed by Thales was caused by friction upon many different types of materials. The entire concept of electron physics and the ensuing electron theory was the outcome of that observation made by Thales in 600 B.C.

Static electricity deals with the action and reaction of electrically charged bodies. How many times have you shuffled across a thick carpet and touched a metal object like a doorknob and drew a sizable spark and experienced an electric shock? As you moved across the carpet, the atoms in your body were accumulating an excessive amount of free electrons from the carpet. When you reached the doorknob and touched it, these excessive free electrons in your body then rushed down through your arm, down through your finger tips onto the doorknob with an accompanying spark.

Static electricity is said to be generated by *friction*. Lightning is a form of static electricity. A thunderstorm generally occurs on days that have been hot and humid. It is caused by the rapid rising of the air. As the warm damp air rushes upward, it is cooled by expansion and the water vapor in it begins to condense into drops of water. This condensation releases heat that warms the air, giving added force to the rising air currents. This condensed moisture then starts to fall. As these drops of water fall through the mass of rapidly rising air, friction with the air tears some of these drops apart into spray or mist. In this process electrical charges are built up, much like your walking across the room on a thick carpet. The water drops falling to the lower levels have been robbed

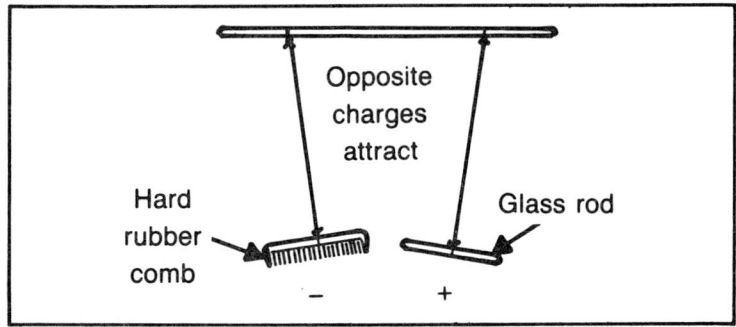

Fig. 1-4. Attraction of unlike charges.

of electrons and take on a positive charge, while the fine spray now rising to the upper levels has picked up an excessive amount of electrons and taken on a negative charge. Sometimes these clouds will accumulate such massive differences of potential that they will discharge from one cloud to another until these differences are neutralized. An ensuing display of spectacular fireworks known as lightning is the result. And it is all caused by friction!

LAW OF ATTRACTION AND REPULSION

If a negatively charged hard rubber comb contacted a positively charged glass rod, electrons would rush from the comb onto the rod (see Fig. 1-4). Figure 1-4 illustrates the electrostatic law that *opposite charges attract each other.* If this negatively charged comb contacted another negatively charged hard rubber comb, the excessive electrons formed on both combs would have a repelling effect (see Fig. 1-5). Figure 1-5 also illustrates the electrostatic law that *like charges repel each other.*

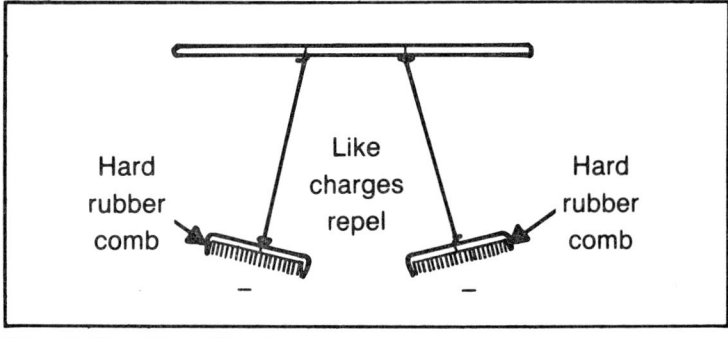

Fig. 1-5. Repulsion of like charges.

ELECTROSTATIC FIELD

This attraction or repulsion of electrically charged bodies is due to an invisible force called an *electrostatic field* that surrounds the charged bodies.

Figure 1-6 represents electrostatic fields existing between unlike and like charges. Each line has a definite direction and is called an *electrostatic line of force*.

Figure 1-7 illustrates the method of representing the lines of force around electrical charges. Note that these lines of force point away from the positive charge and towards the negative charge.

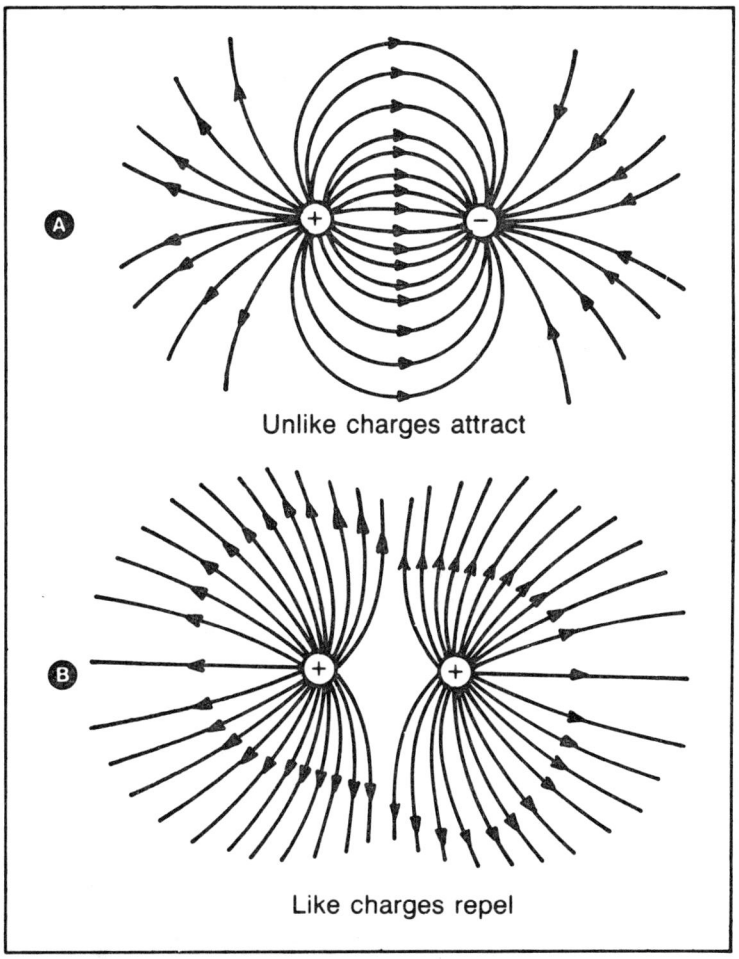

Fig. 1-6. Electrostatic fields.

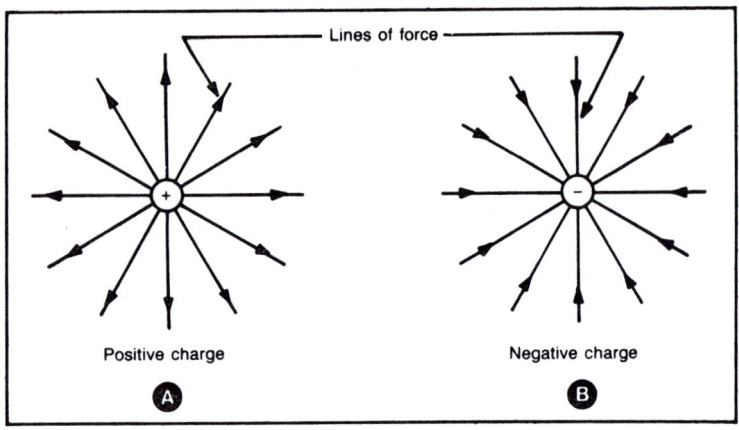

Fig. 1-7. Direction of electrostatic fields.

WORK

In physics, *work* is done when a force acting on a physical body moves that body a definite distance. The amount of work performed can be measured by multiplying the force acting on the body by the distance the body is moved in the direction of that force. If a steady 1-lb force moves a body 1 ft, then 1 foot-pound (ft-lb) of work has been accomplished. Figure 1-8 shows a 15-lb weight raised in the direction of the force a distance of 15 ft. In this operation the direction of the motion is opposite to the force of gravity. The work accomplished is 225 ft-lb, which can be expressed by the formula.

$$W = Fd$$

where:
- W = work accomplished in foot pounds
- F = applied force in pounds
- d = distance in feet that the object is moved

An example of work being accomplished by overcoming friction is shown in Fig. 1-9. When the 200-lb weight is pulled across the floor in the direction of the applied force, work is accomplished by overcoming the resistance of friction between the weight and the floor.

ENERGY

Energy is defined as the ability or capacity to accomplish work. A physical object has energy if it is able to do work.

We saw that the work required to raise a 15-lb weight 15 ft

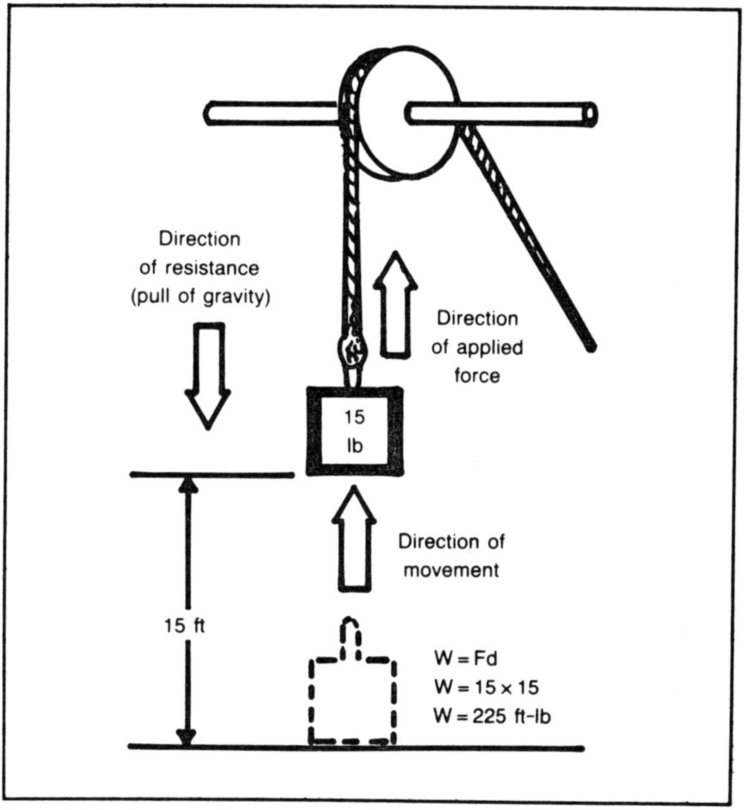

Fig. 1-8. Work accomplished by raising a weight against gravity.

against the force of gravity is 225 ft-lb. This same weight then has the capacity to do 225 ft-lb of work by returning to its original sitting position. So we can say the weight in its raised position possesses energy.

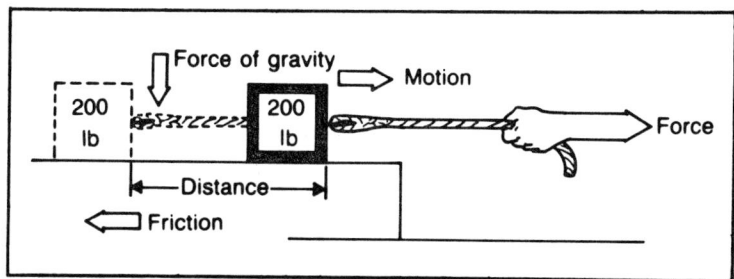

Fig. 1-9. Work accomplished by overcoming friction.

There are two types of energy: potential and kinetic. *Potential* energy is that stored in an object, or that which a body possesses when at rest. Some examples of objects possessing potential energy are coal, gasoline, oil, gunpowder, and a charged electrical battery.

Kinetic energy is due to motion. When an object is moving, it is capable of doing work when it finally comes to rest. Look again at our 15-lb weight raised 15 ft in the air. As it dangles in the air, it has potential energy. If it dropped, this potential energy is transformed into kinetic energy. The farther it falls, the greater is its speed and the greater is the amount of potential energy transformed into kinetic energy.

An important law in physics is the law of conservation of energy: *energy can be changed from one form to another, but it cannot be created or destroyed. Any* form of energy has the potential to do some type of work. Energy is transformed from its original form to some other form in the process of doing work.

When we burn fuel, the heat energy from it can be converted to an electrical energy by driving a turbine, which in turn drives a generator. This electrical energy can now be converted back to heat for cooking and heating. The energy is not destroyed, just transformed.

POWER

Although calculations with electrical power are covered in detail in Chapters 2 and 3, for the sake of continuity we examine the basic concept of power here.

From our look at work and energy we note that the time required to move physical objects was not a factor in determining the amount of work done. We all know that some people can perform work twice as fast as others. Two people can do the same amount of work but at different rates. This rate of doing work is called *power*.

Power is expressed by the formula.

$$P = \frac{W}{t}$$

where:
    P = power in foot-pounds per second
    W = work in foot-pounds
    t = time in seconds

The English scientist James Watt observed that a horse could maintain a 550-ft-lb/sec pace for a reasonable length of time, so he coined the term *one horsepower* (hp).

In electrical and electronic terms, power input and power output are common terms; however, units of electrical power are not the same as mechanical power even though the concept is the same. The electrical unit of power is the *watt* (W), which represents the rate at which energy is transformed. A relationship between electrical and mechanical power is

$$1 \text{ hp} = 746 \text{ W}$$

Table 1-3 illustrates a more comprehensive relationship between various units of power.

EFFICIENCY

The power output from any mechanical engine, mechanism, electric motor, or electrical circuit is always less than the power input. The difference between input power and output power is called *power loss*. In mechanical devices power losses are caused by resistance to electron flow. In both cases power is lost in the form of heat.

Remember the law of conservation of energy: the energy is not lost; it is just transformed into heat and is only lost as the power we desire to do the work.

Efficiency (Eff) is defined as the ratio of power input to power output:

$$\text{Eff in percent} = \frac{\text{Power Output}}{\text{Power Input}} \times 100$$

Table 1-3. Various Relationships between Units of Power.

1 horsepower (hp)	= 550 ft-lb/sec
	= 33,000 ft-lb/min
	= 746 W
	= 0.746 kW
1 watt	= 0.737 ft-lb/sec
1 kilowatt	= 737 ft = lb/sec
1 joule	= 0.738 ft-lb/sec

There is no such thing as a 100-percent efficient device. This formula should prove useful in many applications.

QUIZ

1. In electricity what atomic particle is considered to be the most important?
2. Name the two parts of an atom.
3. What is power?
4. What is lightning?
5. What is a conductor?
6. One horsepower equals _____ watts?
7. What is a molecule?
8. What is an insulator?
9. What are the laws of attraction and repulsion for charged bodies?
10. What is energy?
11. What is the formula for calculating efficiency?
12. What is the definition of matter?
13. What is the best known conductor?
14. Give an example of an object with potential energy.
15. Give an example of a material that is a good insulator.

Chapter 2

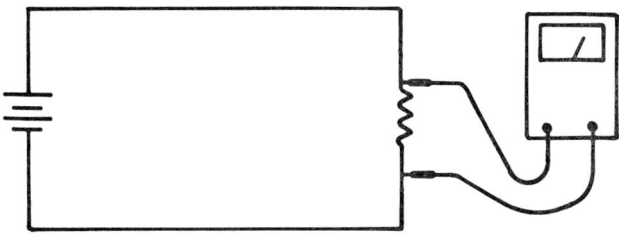

Electrical Properties and Basic Measurement Units

IN CHAPTER 1 YOU WERE INTRODUCED TO THE MORE IMPORtant facets of electron physics. This chapter will enlarge on three fundamental, invisible quantities present in every electrical circuit: voltage, current, and resistance, along with their resultant, power. These quantities are directed and controlled by the proper arrangement of component parts in order to obtain the desired results. Specific terms and symbols are used to identify each component to allow rapid recognition and understanding of its purpose. Things will start taking shape as the simple circuit, schematic symbols, and diagrams are also investigated.

POLARITY

In our discussion on static electricity it was evident that any two differently charged objects provided the force to produce a movement of electrons through a conductor. By knowing which of these two objects has either a negative charge or a positive charge, we can predict the direction of this electron flow. This directional quality is called *polarity*. When a difference of electrical charges exists between two points, one point is said to have a negative (positive) polarity with respect to the other point.

In direct current (dc) electricity, all energy-producing devices have their terminals marked with a minus (−) sign or a plus (+) sign to indicate polarity. You must observe this polarity when making connections in dc circuits to ensure proper equipment operation.

Our first key point to remember is that electron flow (hereafter called *current* or *current flow*) is always from the *negative* terminal of a power source through the circuit and back to the *positive* terminal of that power source. In other words, current always flows from negative to positive. Simply keep in mind that any object having a negative charge has an excessive amount of electrons, and any object having a positive charge has been robbed of electrons. The negative point with this excess of electrons by nature wants to rid itself of them by depositing them at the first available positive point.

It is also true that current flows from the more negative to the more positive point in a circuit.

EMF AND THE VOLT

As we already know, any electrical charge existing between two points has the ability or *potential* to move electrons through a conductor. This potential force is called *electromotive force* (emf). In electrical formulas throughout this book, we will denote it by "E."

In practical terms emf is referred to as electrical pressure, voltage or potential difference. The practical unit of emf is the *volt*. It is named after Allesandro Volta, an Italian scientist who constructed the first battery.

A more scientific definition of the volt is the emf necessary to cause one ampere of current to flow through one ohm of resistance. Right now this definition should mean nothing to you, but later in the chapter we shall define current and resistance.

SOURCES OF EMF

The six principal methods for producing an emf are friction, heat, light, chemical, magnetic, and pressure. Recall from our lesson in electron physics that outside forces on any material can cause electrons to flow. These six methods are the outside forces that can move free electrons through a conductor.

We have already covered static electricity, which is caused by friction. Because electron movement (current flow) must be maintained to do any appreciable amount of work, such as operate light bulbs and electric motors, producing an emf by friction is useless and can be dispensed with.

Heat

If two dissimilar metals are joined and the junction heated, electron movement occurs and will continue until the heat is removed. This method is called the *thermoelectric effect.*

A practical device that has evolved from the thermoelectric effect is the thermocouple. A *thermocouple* is a temperature-sensing element that creates current flow in proportion to the temperature at the element (junction). For example, to make an automobile temperature gauge, you need only provide a path for this current flow by inserting a specially calibrated ammeter that gives a readout in degrees Fahrenheit (F) or Celsius (C).

Light

When exposed to light rays, certain materials cause a movement of electrons and consequently can become a source of emf. This method called the *photoelectric effect,* is responsible for numerous applications where very small current values are required; one well-known use is the photographic light meter.

Today, solar-powered devices are quite important, and the photovoltaic cell will eventually play an even bigger part.

Figure 2-1 illustrates the photoelectric effect. Figure 2-2 shows one possible use of solar power as a source of emf.

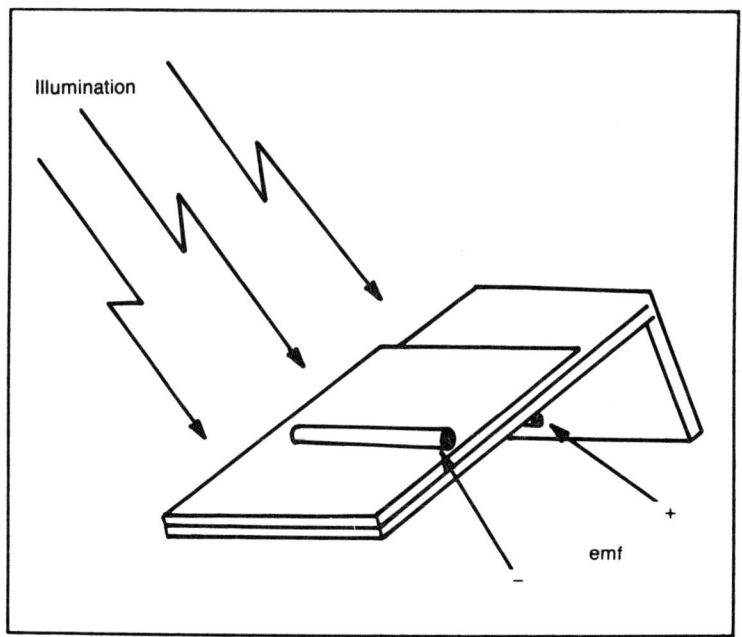

Fig. 2-1. A selenium photo-cell illustrating the photoelectric effect.

Fig. 2-2. A 1912 Baker Electric, the first solar-powered car in history, uses 10,640 International Rectifier solar cells to convert sunlight directly into electrical energy for charging its storage batteries. This dramatic demonstration emphasizes the possibilities of harnessing sunlight to provide power not only for automobiles, but for a variety of electrically driven devices. (Courtesy International Rectifier Corp.)

Chemical

Without a doubt, chemical means are one of the most widely used and important methods for creating an emf. One reason for this popularity is that chemical action produces an emf large enough to supply a usable amount of current flow for general applications.

Chemical energy may be converted to electrical energy by means of cells. These cells are like a pump that removes electrons from one end and piles them up at the other end. This creates a potential difference. We know these devices as *batteries*. Batteries are covered in detail in Chapter 5. Figure 2-3 illustrates a cross-sectional view of a dry cell battery. When two dissimilar substances like zinc and carbon are placed in a salt solution (e.g., ammonium chloride), a chemical action takes place. Electrons are torn from the carbon rod, making it positively charged, and piled up on the zinc shell, making the entire case negatively charged.

Magnetic

Magnetic means are the most common and useful way to create an emf. An early experimenter named Lenz discovered that if a conductor moves through a magnetic field, then a potential difference is set up between the ends of that conductor.

Figure 2-4A shows a coil consisting of many turns of wire wound about a bakelite tube with a bar magnet hovering above it.

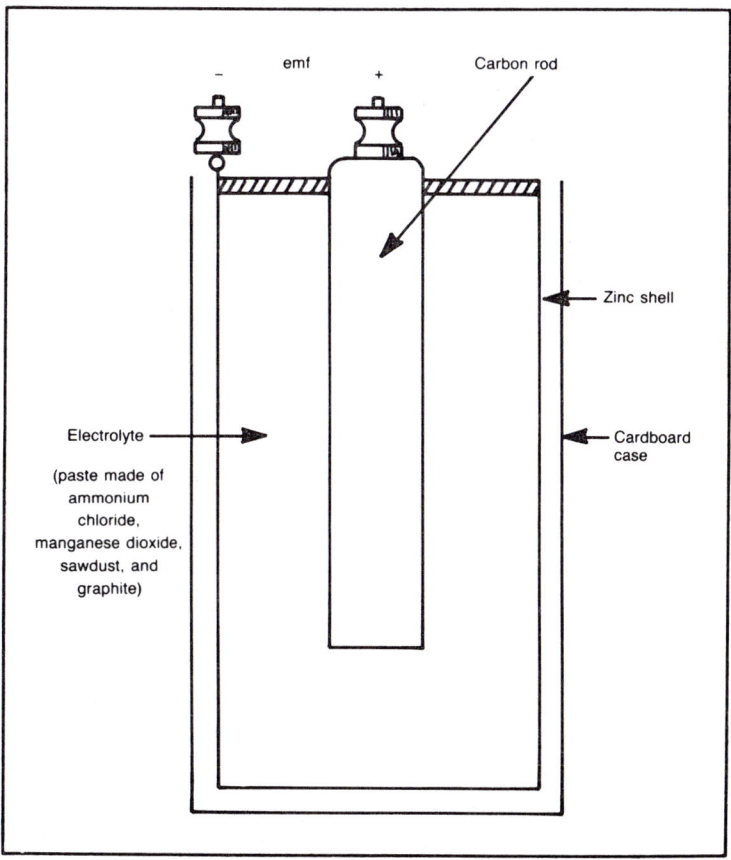

Fig. 2-3. Cross-sectional view of a dry cell battery.

21

In this position there is no relative movement, so there is no emf being induced. Shifting down to Figure 2-4B, we see that the magnet has been plunged into the center of the tube. It is only during this plunging movement that an emf is induced. When the magnet comes to rest and there is no movement, there is also no emf.

This is the principle of the electric generator.

Note that as the strength of the magnetic field increases, the number of lines of force cut during a specified time also increases. Also, as the relative motion between this field and the conductor increases, the number of lines of force cut during a specified time increases. Thus, the strength of the induced emf depends upon the number of lines of force cut per second. By definition, if 100,000,000 lines of force are cut per second, an induced emf of one volt will be produced. We may increase the emf by increasing the strength of the magnetic field, the speed of the relative motion, or the number of conductors or turns of wire cutting the field.

Pressure

Under pressure certain crystals create an emf. Conversely, by applying an emf to these crystals, (e.g., quartz) they will convert this electrical energy back to mechanical energy. This is known as the *piezoelectric effect.*

A practical example of this method is the phonograph. The needle is attached to a crystal element on the pickup arm. As the record revolves around the turntable, the needle rides the preformed grooves of the record and applies pressure variations that are felt at the crystal element. The crystal element then produces an emf, with the corresponding electrical variations being in exact proportion to the mechanical variations.

As you can see, all six ways of producing an emf have their own particular applications, but magnetic and chemical means are the most useful for general applications.

CURRENT FLOW AND THE AMPERE

We have now reached the subject that is considered the "meat" of electricity—current flow. When teaching basic electronics at the U. S. Navy Radioman School, I often advised students faced with a difficult circuit analysis problem to imagine themselves as current. This always seemed to put them on the right track.

An electrical current has already been described as a flow of

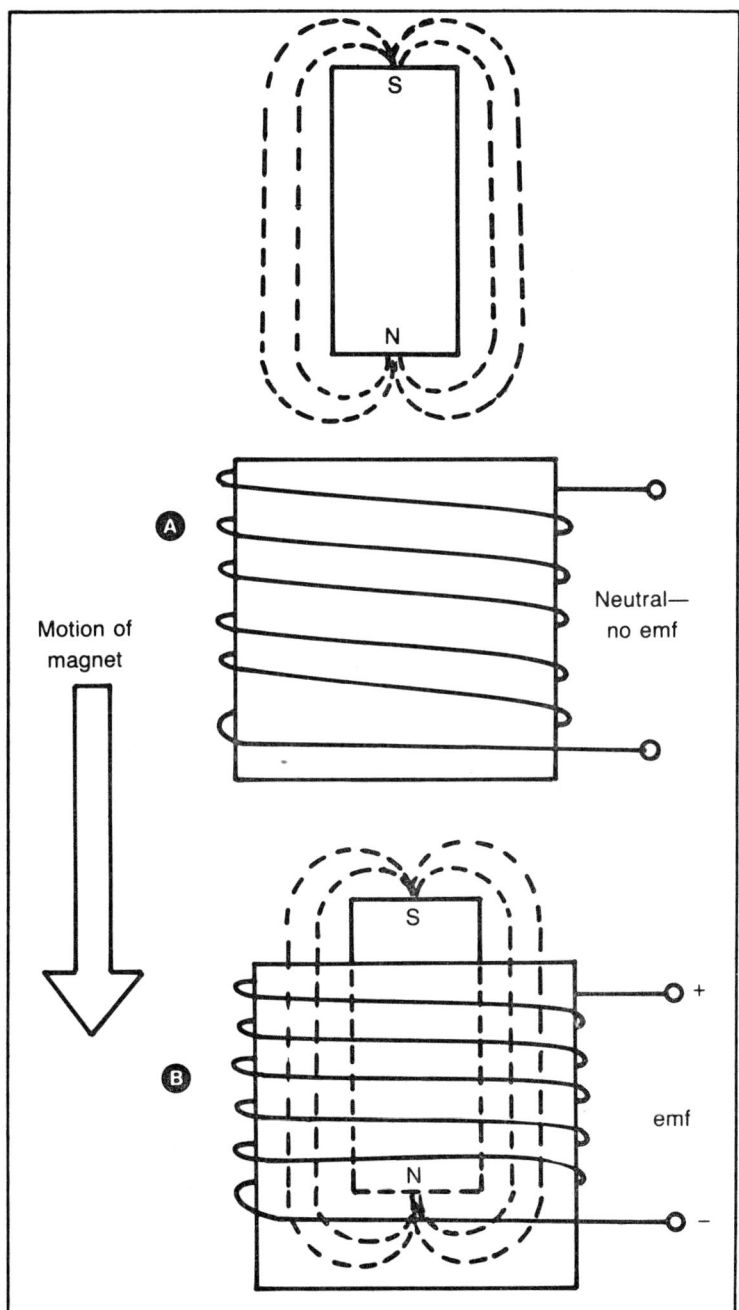

Fig. 2-4. Generating an emf by magnetic induction.

electrons through a conductor (wire) from a point of negative potential to a point of positive potential.

The basic unit of current flow is the ampere (A), named after the French scientist Andre Ampere, whose work with electromagnetism led to the measurement of current. A current flow of one *Ampere* is said to be flowing when 6,280,000,000,000,000,000 electrons pass one point in one second!

Electrical current is measured by an ammeter. The actual speed at which these electrons move through a conductor is relatively slow; however, the disturbance that causes this movement through the conductor (wire) takes place at a speed approaching the speed of light (186,000 mi/sec). This action is illustrated in Fig. 2-5. Assume the hollow tube is 186,000 mi in length and completely filled with balls; if a ball is added at one end, a ball at the other end is forced out one second later. Even though each ball moves slowly and only a very short distance, the disturbance itself is transmitted almost instantaneously through the entire length of the tube.

In electrical and electronic circuits, the letter "I" is used to indicate current. This is derived from the word "intensity."

Current is classified into two general types: direct current and alternating current. They are equally important in almost every kind of electrical and electronic devices. As a general rule, ac is used to supply power to a piece of equipment, and dc is used to operate the internal electronic components. In this book we are concerned only with the principles, laws, and formulas of direct current.

Direct current flows in only one direction—always negative to positive. The voltage sources that produce direct current are commonly referred to as dc power sources or dc power supplies.

RESISTANCE AND THE OHM

As current flows through any type of material, it encounters

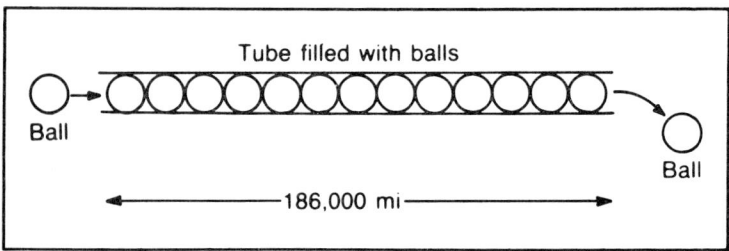

Fig. 2-5. Illustration of electron motion through a conductor.

24

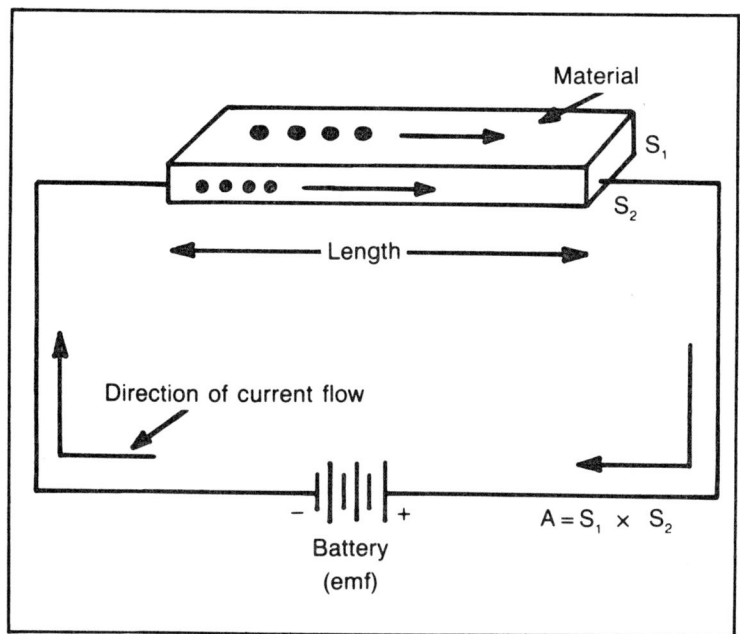

Fig. 2-6. Resistivity of a material.

resistance. This electrical resistance is based on the natural resistivity of the material it is flowing through.

Georg Simon Ohm, a German physicist, observed that with a fixed emf the amount of current through any material depended on the type of material and the physical dimensions of that material. Fig. 2-6 illustrates this observation. The material shown is rectangular for explanation purposes and assumed to be connected across a battery (emf) with connections offering zero resistance.

If a constant emf is applied across the material to cause current to flow, we can calculate resistance of this material by the formula

$$R = p \times \frac{l}{A}$$

where: R = resistance in ohms
l = length in centimeters
A = cross-sectional area in square centimeters
p = resistivity (see Table 2-1)

Substituting these values into the equation

Table 2-1. Resistivities of Materials (in ohms per circular mil-foot at 20°C).

Material	Resistivity
Silver	9.56
Copper	10.37
Aluminum	17.0
Tungsten	34.0
Brass	42.0
Nickel	60.0
Iron	61.0
Platinum	66.0
Manganin	264.0
Constantan	295.0
Cast iron	436.0
Nichrome	675.0
Carbon	22,000.0

$$R = p \times \frac{1}{1}$$

Therefore, $R = p$. Refer to Table 2-1 for the type of material and resistivity. If the material in our example were carbon, our answer would be 22,000 ohms.

The basic unit of electrical resistance is the *ohm*, designated by the symbol Ω. You should remember that resistance is present in the components and wires of every electrical circuit. The amount of resistance offered by wires is very small compared to the overall circuit resistance, so in practical work it is considered to be zero.

THE SIMPLE CIRCUIT

An electrical circuit is simply a path (or paths) for current to flow. In Fig. 2-7 the schematic symbol for a battery is shown. Current has no place to flow. The battery as a source of emf just sits there with a difference in potential existing. By attaching a wire from the negative terminal to the positive terminal of the battery, (Fig. 2-8) we provide a path for current to flow. This is the simple

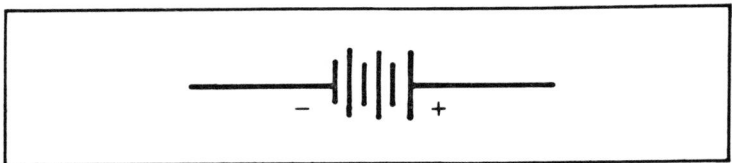

Fig. 2-7. Schematic symbol for a battery as a source of emf (chemical).

Fig. 2-8. A path for current to flow (the simple circuit).

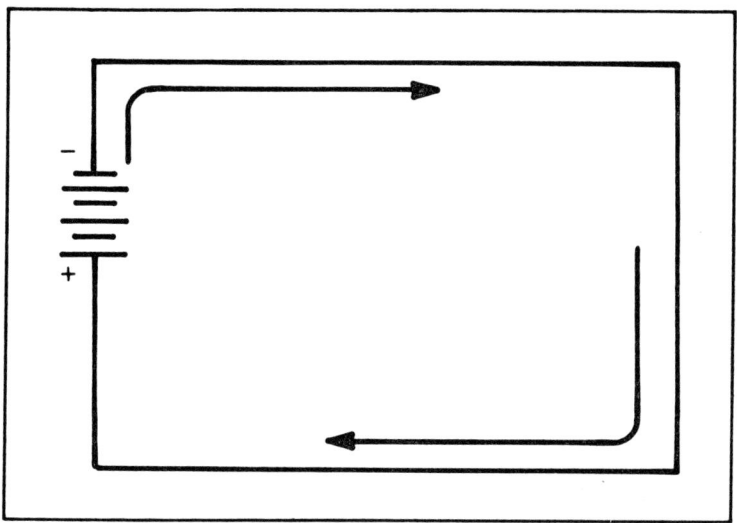

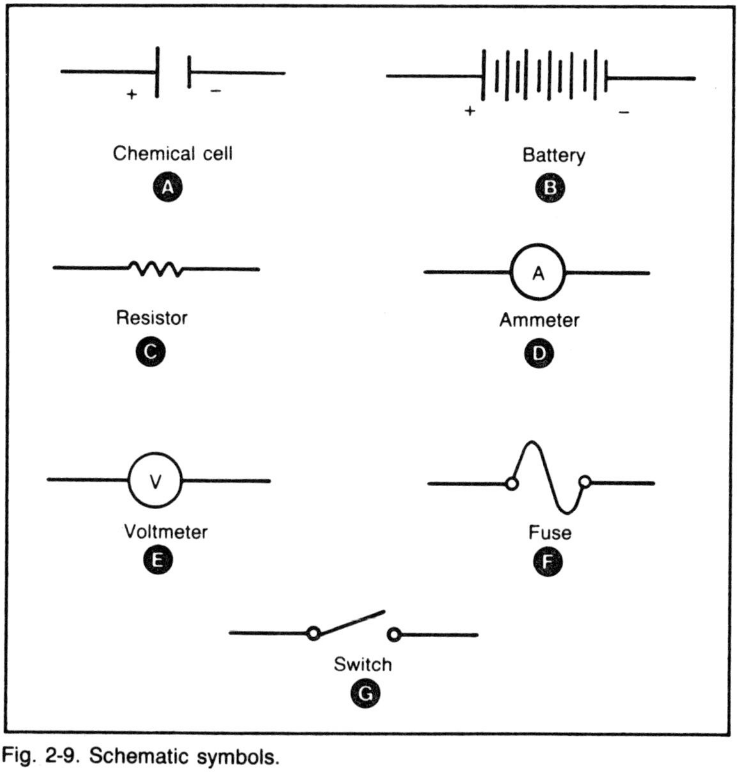

Fig. 2-9. Schematic symbols.

27

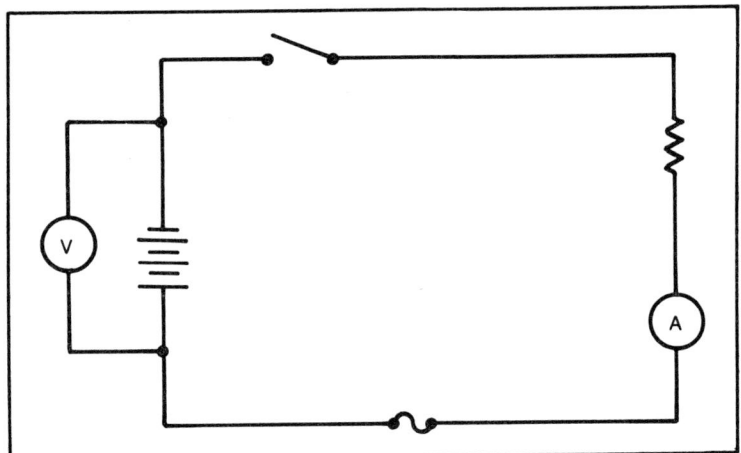

Fig. 2-10. Schematic diagram of a practical electric circuit.

circuit. Unfortunately, this simple circuit won't do much except exhaust the battery, and if the size of the wire isn't adequate to handle the current flowing through it will heat up and break, thereby leaving an *open circuit* or no circuit at all.

A practical circuit limits the current flow so it performs useful operating work, such as motors, electronic devices, lamps, and the like.

BASIC SCHEMATIC SYMBOLS

To get ready for the practical work in the next chapter, we need to become familiar with a few important schematic symbols that will be used in the practical circuits.

Figure 2-9 shows the schematic symbols for a chemical cell (A); a battery—several cells (B); a resistor (C) (see Chapter 6); an ammeter (D); a voltmeter (E); a fuse (F), which is a protective device designed to heat up and open the circuit ("blow") if the current flow becomes too high; and a switch (G), which is a device for opening and closing a circuit (i.e., turning it on or off). Figure 2-10 incorporates all of these devices into a simple circuit. A more comprehensive listing of schematic symbols is in Appendix G.

QUIZ

1. Electrical current flows from _____ to _____.

2. The volt is the basic unit of _____.
3. An emf can be produced in _____ ways.
4. What are the two most common methods for producing an emf?
5. The basic unit of current flow is the _____.
6. What is the basic unit of resistance?
7. Does copper or carbon offer the most electrical resistance to current flow?
8. E is the symbol for _____.
9. I is the symbol for _____.
10. R is the symbol for _____.
11. Draw the schematic symbol for a battery.
12. Draw the schematic symbol for a fuse.
13. Draw the schematic symbol for a switch.
14. Draw the schematic symbol for a resistor.
15. Schematically, draw an electrical circuit diagram containing a battery, a switch, and a resistor.

Chapter 3

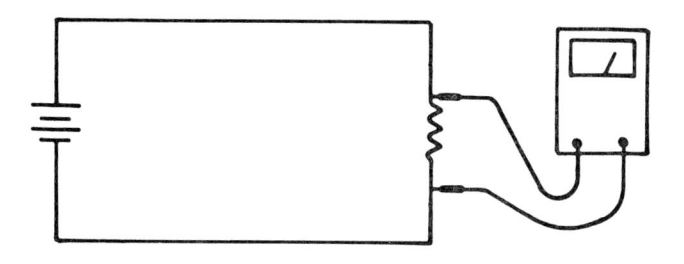

Dc Circuit Analysis

IN THIS CHAPTER WE COVER THE MOST PRACTICAL WORK OF the entire book. You will be introduced to the very important relationships between voltage (emf), current, resistance, and power.

The laws and formulas in this chapter are the absolute foundation for serious entry into the electrical/electronic industry, and they should be *memorized*.

MATHEMATICS FOR ELECTRICITY

The only mathematics you need are simple multiplication, division, addition, and subtraction. However, when measuring resistance, capacitance, and inductance in electronic circuits, you will be using very large numbers, such as 4,000,000,000 and very small numbers, such as 0.00000150. For example, the megohm (MΩ), a practical unit of resistance, equals, 1,000,000 Ω, and the picofarad (pF), a practical unit of capacitance, equals 1/1,000,000,000,000 farad (F). Such numbers are very awkward to work with. An efficient and easy way to express large numbers is by "the powers of 10," sometimes called scientific notation or engineering shorthand.

Powers of 10

In powers of 10, the number 1,000,000 is written 10^6 and read "10 to the sixth power"; 1/1,000,000,000,000 is written 10^{-12} and read "10 to the minus 12th power." The numbers 6 and −12 are called *exponents*; they show how many tens must be multiplied together to equal 1,000,000 and 1/1,000,000,000,000, respectively. Table 3-1 illustrates several different powers of 10 and the numbers they equal.

For positive (negative) exponents of 10, the power indicates the number of places that the decimal point must be moved to the *right* (*left*), starting from its position following the 1. For example,

$10^8 = 100,000,000$ Move decimal point 8 places to the right from 1

$10^{-3} = 0.001$ Move decimal point 3 places to the left from 1

Any number can be expressed as a number between 1 and 10 multiplied by a power of 10. For example, in Chapter 2 we said that 1 A equalled 6,280,000,000,000,000,000 electrons passing a point in one second. This very large figure can be expressed very briefly in powers of 10 as 6.28×10^{18}.

Take a moment to check your understanding by doing the following exercises. Answers to these problems are at the end of the chapter.

(a) Express in powers of 10	(b) Express in numerals
550,000	1.7×10^3
765,000	8.6×10^{-3}
856	8.8×10^2
8390	4.5×10^7
45.1	2.8×10^6
1,875,000	7.9×10^{-4}
55,500	4×10^5
182	3.4×10^3
296,000	4.11×10^3
15,900,000	9.71×10^6
8,510	9.6×10^{-4}
11,600	2×10^{-3}
5500	
3,290,000	
23.7	
0.0001	
0.0000003	
0.001045	
0.0038	
0.00042	

Table 3-1. Powers of Ten.

Positive Exponents		Negative Exponents	
Number	Power of Ten	Number	Power of Ten
1	10^0	1	10^0
10	10^1	0.1	10^{-1}
100	10^2	0.01	10^{-2}
1000	10^3	0.001	10^{-3}
10000	10^4	0.0001	10^{-4}
100000	10^5	0.00001	10^{-5}
1000000	10^6	0.000001	10^{-6}

Multiplication by Powers of 10

For the simple problem

$$10^3 \times 10^2$$

by ordinary methods the product is

$$10^3 \times 10^2 = 1000 \times 100 = 100{,}000$$

In powers of 10, however, we simply *add* the exponents:

$$10^3 \times 10^2 = 10^{3+2} = 10^5 = 100{,}000$$

This is also for negative exponents:

$$10^{-3} \times 10^{-2} = 0.001 \times 0.01 = 0.00001$$
$$10^{-3} \times 10^{-2} = 10^{-5} = 0.00001$$

Division by Powers of 10

The problem

$$10^5 \div 10^2$$

can be written

$$100{,}000 \div 100 = 1000$$

In powers of 10, however, we need only subtract the exponents:

$$\frac{10^5}{10^2} = 10^{5-2} = 10^3 = 1000$$

PREFIXES

In electrical problems, very wide ranges of values of voltage, resistance, capacitance, power, and inductance are often used. For example, capacitance and inductance values may be on the order of millionths or thousandths of their basic units. There are also times when thousands of volts or millions of watts are in order. Prefixes are therefore used for these values. Table 3-2 shows the most common prefixes used, the magnitudes they represent and their equivalent powers of 10 expressions. It is important that you become thoroughly familiar with these prefixes.

In most electrical equations, values of voltage, current, and other quantities must be converted back to their basic units of amperes, volts, and so on before they can be used in the equations. To convert from one unit to another, you multiply the unit to be converted by the appropriate power of 10.

OHM'S LAW

In 1827, Georg Simon Ohm published an article entitled "The

Table 3-2. Commonly Used Prefixes for Electrical Units.

Prefix	Magnitude	Power of 10	Electrical Unit
Mega or Meg (M)	1,000,000	10^6	Megohm
Kilo #(k)	1,000	10^3	Kilovolt Kilowatt
Milli #(m)	$\frac{1}{1,000}$	10^{-3}	Millivolt Milliampere Millihenry
Micro #(μ)	$\frac{1}{1,000,000}$	10^{-6}	Microvolt Microampere Microhenry
Pico #(p)	$\frac{1}{1,000,000,000,000}$	10^{-12}	Picofarad

Galvanic Circuit, Mathematically Expressed." Today, the contents of that article are known simply as *Ohm's Law*. Ohm's Law is one of the most fundamental and important laws of electricity.

This law, formed by carefully observing the relationship between voltage, current, and resistance, states that the current flowing in an electrical circuit is directly proportional to the voltage and inversely proportional to the resistance.

Expressed mathematically it is

$$I = \frac{E}{R}$$

where: I = current in amperes
E = voltage in volts
R = resistance in ohms

Figure 3-1 illustrates Ohm's Law. Assume that the voltmeter measures 15 V across the battery, and the resistor has a resistance of 15 Ω. We want to compute the value of current flowing in the current. Using Ohm's Law, we can calculate circuit current by inserting the value of voltage E and the total resistance R into the formula and then solving the equation for I.

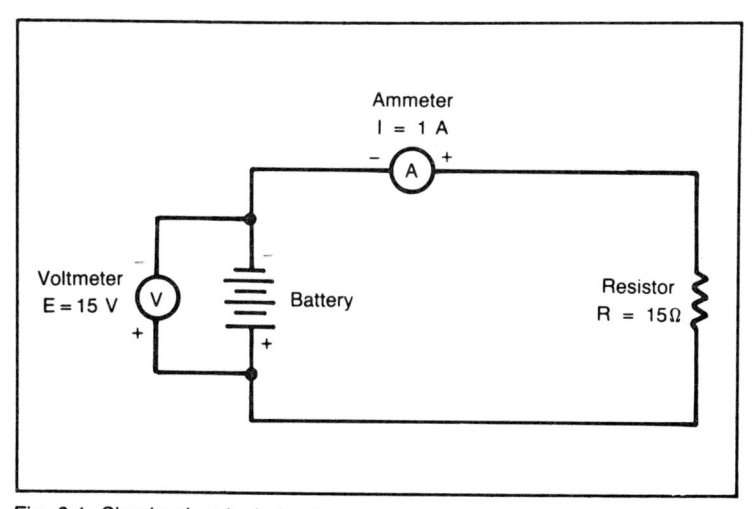

Fig. 3-1. Simple electrical circuit.

$$I = \frac{E}{R}$$

$$= \frac{15 \text{ V}}{15 \text{ Ω}}$$

$$= 1 \text{ A}$$

Thus, the current in the circuit is the same as that indicated on the ammeter. You can use Ohm's Law to calculate circuit current when it is not desirable to insert an ammeter into the circuit.

We can derive two additional expressions of Ohm's Law from this equation:

$I = \dfrac{E}{R}$ (current is equal to voltage divided by resistance)

$R = \dfrac{E}{I}$ (resistance is equal to voltage divided by current)

$E = I \times R$ (voltage is equal to current multiplied by resistance)

Now for any electrical circuit containing two known values, we can easily solve for the unknown values. These expressions are so important that you must understand them and memorized them.

A simple aid for remembering Ohm's Law is illustrated in Fig. 3-2. This is known as the circle method, and it is quite easy to remember. If you cover E with your fingertip, you see that I and R are side by side, indicating that you need to multiply to find E. If you cover the I, you see that E is over R, indicating that you need to divide the two to determine I. Finally, by covering R, you can see that E is over I, which indicates that you must divide the two to find R.

Effect of Increasing Voltage

Figure 3-3 is a basic circuit with a table indicating the circuit quantities for two different conditions. In both conditions resistance is kept constant and voltage is increased. Comparing condition 2

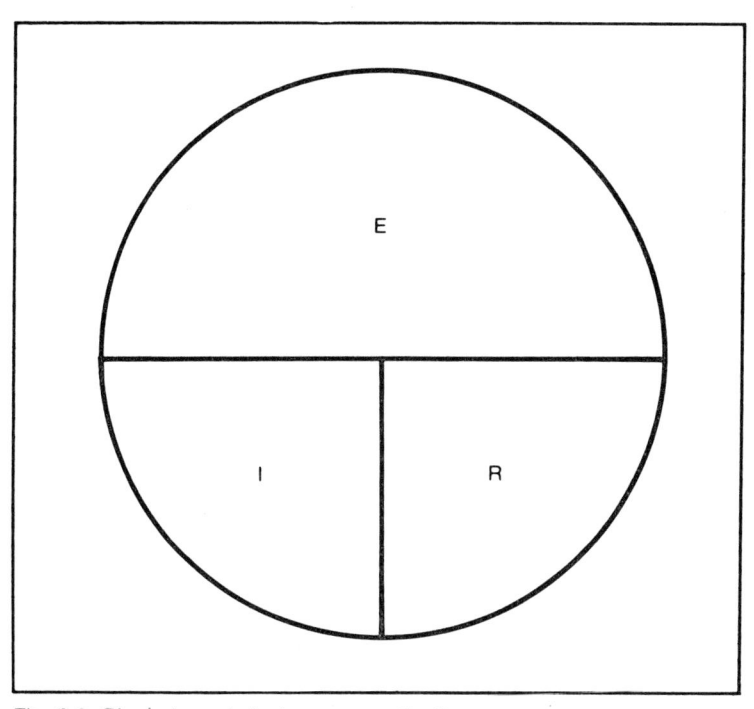

Fig. 3-2. Ohm's Law circle (memory method).

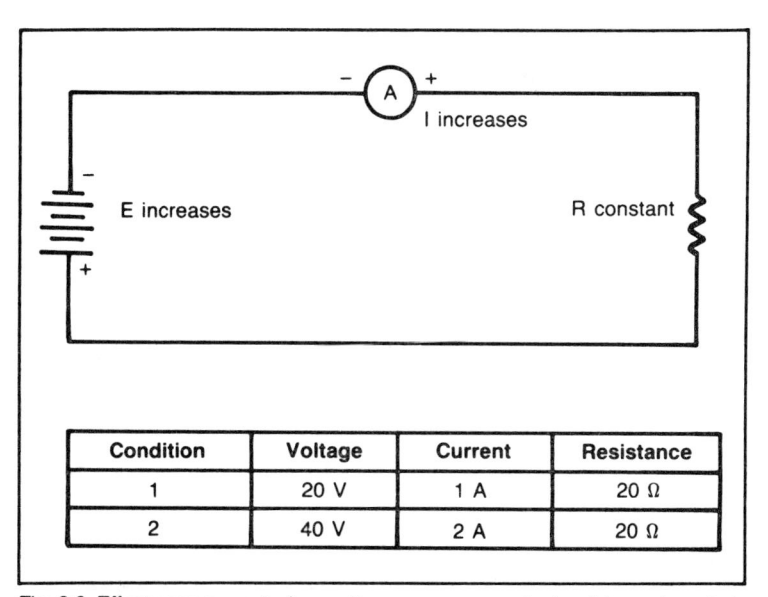

Condition	Voltage	Current	Resistance
1	20 V	1 A	20 Ω
2	40 V	2 A	20 Ω

Fig. 3-3. Effect upon current when voltage across a constant resistance is varied.

to condition 1, we see that doubling the voltage results in doubling the current flow when resistance remains constant. From this analysis it is evident that if the resistance of any circuit is kept constant and the voltage is increased, the current will increase in direct proportion to that increase in voltage.

Effect of Increasing Resistance

Figure 3-4 is similar to Fig. 3-3, except that resistance is varied and voltage remains constant. Referring to the table of Fig. 3-4 and comparing condition 2 to condition 1, we see that with the same voltage applied, doubling the resistance in the circuit reduces the current to one-half the original amount. We can conclude that if in any circuit the voltage is kept constant and the resistance is increased, the current will decrease by an amount proportional to that increase.

ELECTRICAL CIRCUITS

An electrical circuit is a path or group of interconnected paths capable of carrying an electrical current. There are three basic types of circuits: series, parallel, and series-parallel. These circuits may range from very simple networks to extremely complex networks.

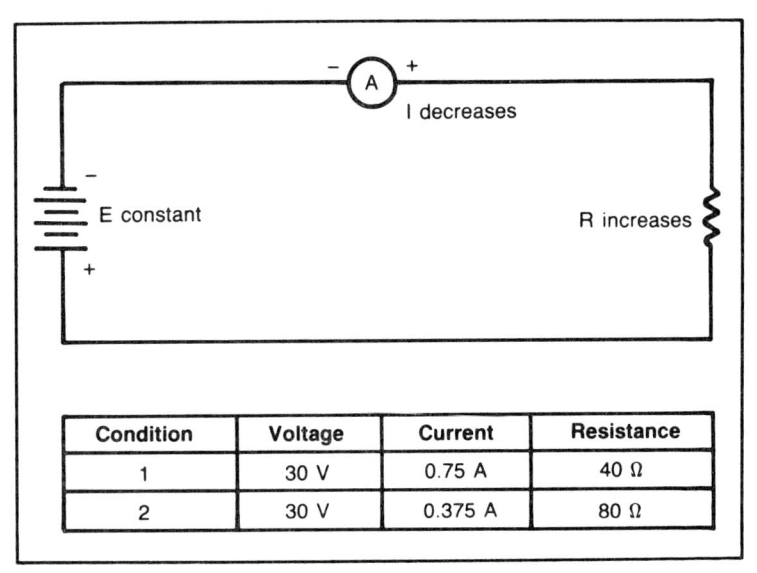

Condition	Voltage	Current	Resistance
1	30 V	0.75 A	40 Ω
2	30 V	0.375 A	80 Ω

Fig. 3-4. Effect upon current when resistance is increased and voltage remains constant.

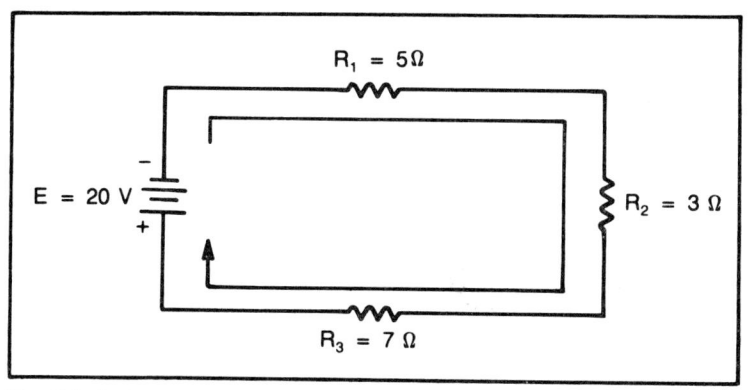

Fig. 3-5. Series circuit.

It is very important that you become familiar with the characteristics of each type of circuit because each has its own set of laws and formulas. Then by Ohm's Law circuit operation can be understood and unknown circuit values can be obtained.

SERIES CIRCUITS

Current

A *series* circuit is one in which there is only one path for current to flow. In a series current (see Fig. 3-5) the same amount of current is flowing in all parts of the circuit. In other words, the current flowing through R_1 is the same as the current flowing through R_2 is the same as the current flowing through R_3 and is equal to the current supplied by the battery.

The two most important facts to remember about a series circuit are that

1. There is only one path in which current can flow.
2. Current is the same in all parts of the circuit.

Resistance

In a series circuit each resistor opposes the flow of current. The total opposition is the sum of the resistances of the individual resistors:

$$R_t = R_1 + R_2 + R_3 + \ldots$$

You can use Ohm's Law to find the total resistance of any number of resistors connected in series. This total is always equal

39

to the sum of the individual resistances. The total resistance of the circuit in Fig. 3-5 is

$$R_t = 5\ \Omega + 3\ \Omega + 7\ \Omega$$
$$= 15\ \Omega$$

These three resistors offer the same opposition to current flow as a 15 = Ω resistor. To find current flow, we simply apply Ohm's Law:

$$I = \frac{E}{R}$$

$$= \frac{20}{15}$$

$$= 1.33\ A$$

A more difficult example is the circuit in Fig. 3-6. To find the total current in the circuit, we must first find the total resistance. We first convert all resistance values to their basic unit of ohms: $R_1 = 3300\ \Omega$, $R_2 = 1200\ \Omega$, $R_3 = 1800\ \Omega$, $R_4 = 1500\ \Omega$, and $R_5 = 1200\ \Omega$. Because

$$R_t = R_1 + R_2 + R_3 + R_4 + R_5$$

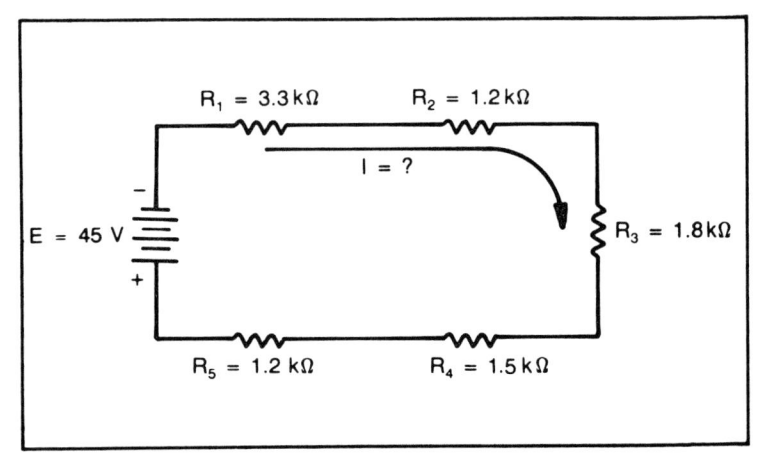

Fig. 3-6. Finding current in a series circuit.

then

$$R_t = 3300 + 1200 + 1800 + 1500 + 1200$$
$$= 9000 \ \Omega$$

Now, knowing R_t and E, we can solve for I:

$$I = \frac{E}{R}$$

$$= \frac{45}{9000}$$

$$= 0.005 \ A$$

It is common practice to use the milliampere (mA) (refer to Table 3-2). Multiply by 1000 or simply move the decimal point three places to the right. The total current is then 5 mA.

Voltage

A *voltage drop* is the amount of emf being used up as current flows through the circuit. When current flows through a resistor, a voltage drop is created. Refer to Fig. 3-7. If you connected a voltmeter to points A and B, it would read 5 V because 1 A of current is flowing through a 5-Ω resistor. Use Ohm's Law to prove this:

$$E = I \times R$$
$$= 1 \times 5$$
$$= 5 \ V$$

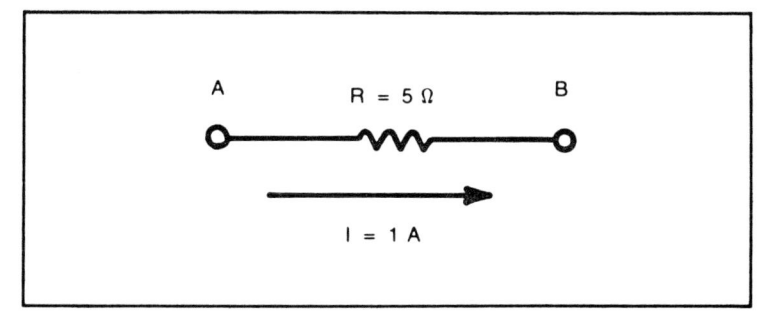

Fig. 3-7. Voltage drop across a resistor.

Fig. 3-8. $E_t = E_1 + E_2 + E_3$.

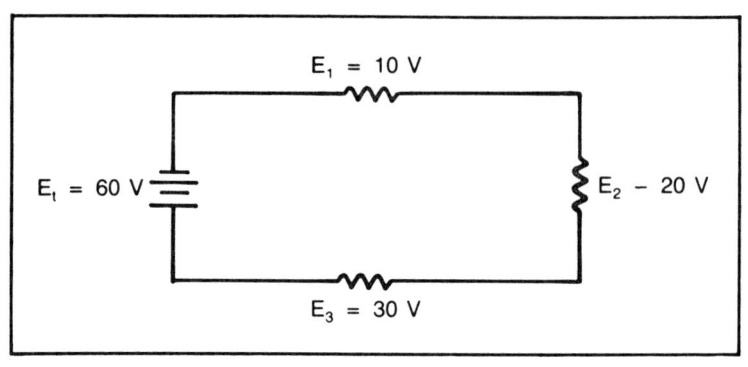

The total voltage is a series circuit is the sum of all individual voltage drops. These voltage drops always equal the source voltage. In symbols this is

$$E_t = E_1 + E_2 + E_3 + \ldots$$

See Fig. 3-8.

To check our understanding of series circuits, we will solve the circuit in Fig. 3-9, which at first glance might seem impossible. Is there enough information provided? Yes. Looking at the individual resistors, we see that R_2, which is 40 Ω, has created a voltage drop of 20 V. With these two known values we can get started. From Ohm's Law

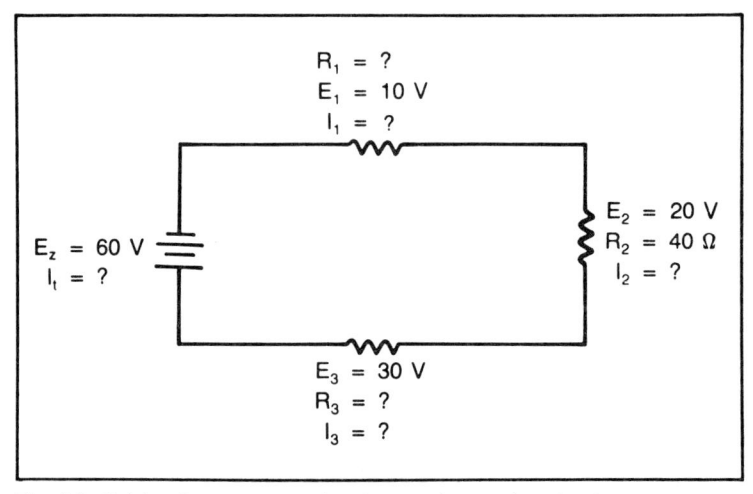

Fig. 3-9. Solving for current and resistance in a series circuit.

$$I = \frac{R_2}{E_2}$$
$$= \frac{20 \text{ V}}{40 \text{ }\Omega}$$
$$= 0.5 \text{ A}$$

Because current is the same throughout a series circuit, we know that

$$I_1 = I_3 = I_t = 0.5 \text{ A}$$

Next, we can solve for R_1 by using Ohm's Law.

$$R_1 = \frac{E_1}{I_1}$$
$$= \frac{10 \text{ V}}{0.5 \text{ A}}$$
$$= 20 \text{ }\Omega$$

The solving for R_3 the same way gives

$$R_3 = \frac{E_3}{I_3}$$
$$= \frac{30 \text{ V}}{0.5 \text{ A}}$$
$$= 60 \text{ }\Omega$$

Now we can determine the only remaining value, R_t:

$$\begin{aligned} R_t &= R_1 + R_2 + R_3 \\ &= 20 \text{ }\Omega + 40 \text{ }\Omega + 60 \text{ }\Omega \\ &= 120 \text{ }\Omega \end{aligned}$$

We can check our work by Ohm's Law

$$E_t = I_t \times R_t$$
$$= 0.5 \text{ A} \times 120 \text{ }\Omega$$
$$= 60 \text{ V}$$

Power

If the emf in volts is multiplied by the current in amperes, we get the value of power (P) dissipated in watts (W).

$$P = E \times I$$

Example. For the circuit in Fig. 3-10, calculate the power dissipated by the light bulb.

$$P = I \times E$$
$$P = 0.5 \text{ A} \times 110 \text{ V}$$
$$P = 55 \text{ W}$$

In this example we can say that an emf or electrical pressure of 110 V is causing 0.5 A to flow in the circuit. This is causing 55 W of energy to be used (dissipated) in the light bulb. This 55 W of energy is being dissipated in the form of light and heat. The handy wheel in Fig. 3-11 can be used to remember the unknown values in figuring power. Use it just as you did the Ohm's Law wheel in Fig. 3-2.

PARALLEL CIRCUITS

A *parallel* circuit consists of connections of components that provide two or more paths in which current can flow. An example of a simple two-branch parallel circuit is shown in Fig. 3-12. Trac-

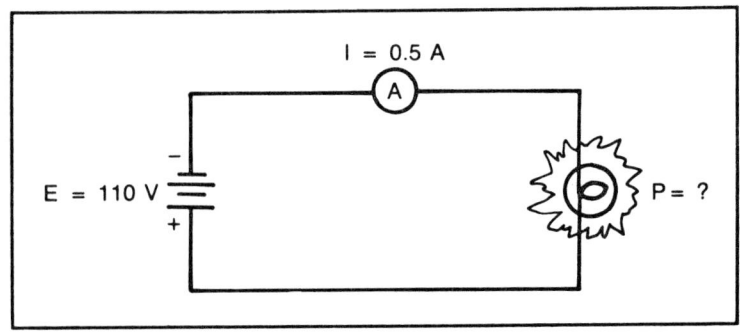

Fig. 3-10. Solving for power in a series circuit.

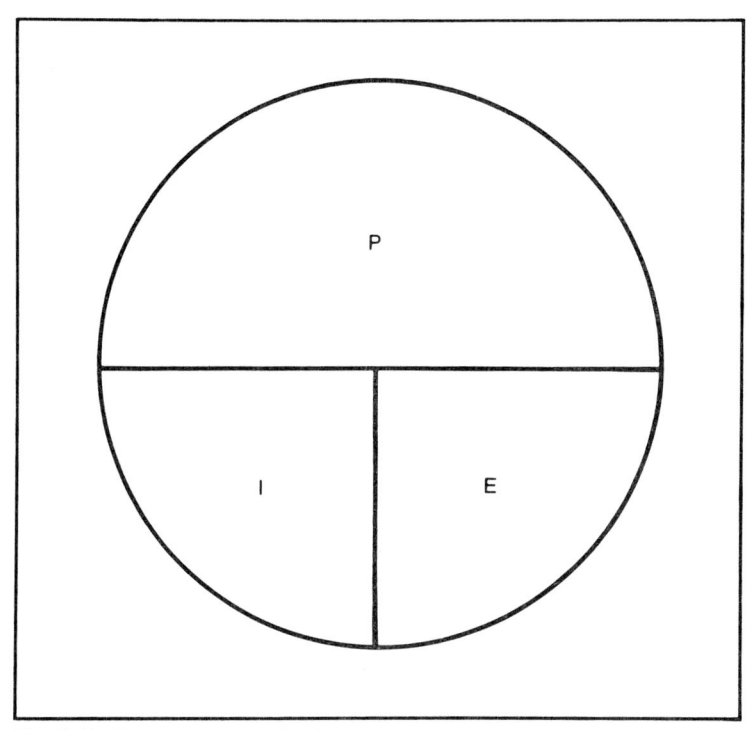

Fig. 3-11. Power memory wheel.

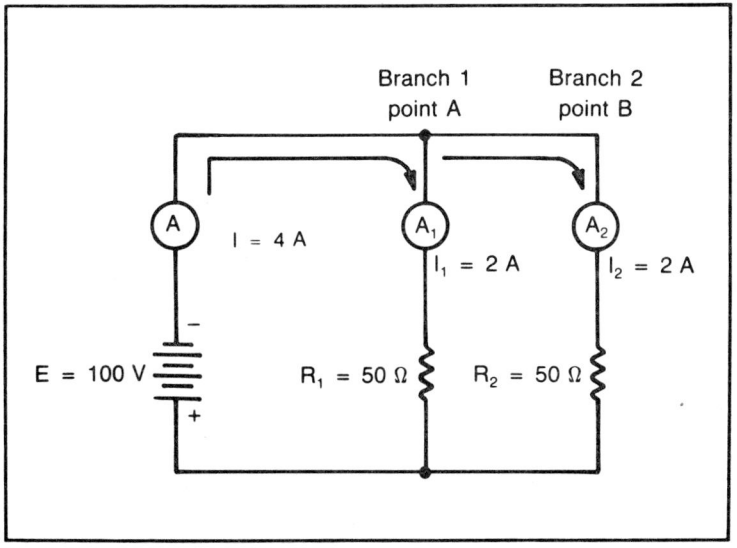

Fig. 3-12. Simple parallel circuit.

45

ing current flow through this circuit, we see that it leaves the negative terminal of the source battery through the ammeter to point A. Here the current can take two paths to return to the positive terminal of the battery. One path is through ammeter A_1 and resistor R_1; the other is through ammeter A_2 and resistor R_2.

From the diagram, you can see that the current measured by ammeters A_1 and A_2 is not same as the current measured by the ammeter near the battery. The current measured by this ammeter is the sum of the currents measured by A_1 and A_2.

From this observation our first law and electrical formula for parallel circuits is derived. In a parallel circuit the total current is equal to the sum of the currents in the individual branches:

$$I_t = I_1 + I_2 + I_3 + \ldots$$

Figure 3-13 shows a parallel circuit composed of a 24-V battery and two parallel resistors each having 6 Ω of resistance. In this circuit a voltmeter is used to measure the battery voltage. If you removed the voltmeter from the battery and connected it across points A and B, the voltmeter would read the battery voltage of 24 V.

Tracing the current from the negative terminal of the battery through the circuit and back to the positive terminal of the battery, we see that the total current passes through the ammeter on its way to point A. At point A the current sees two paths to take on its return to the positive terminal of the battery. At this point it is a good idea for you to mull over a simple scientific point: *current always takes the path of least resistance.* Looking at Fig. 3-13 again, we see that branch 1 offers 6 Ω of resistance, and branch 2 also

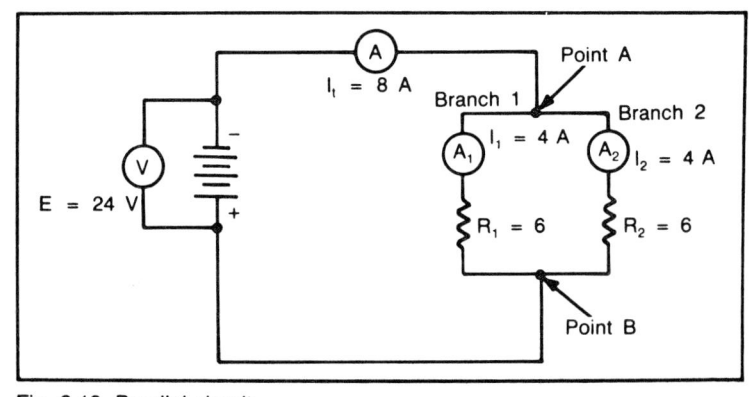

Fig. 3-13. Parallel circuit.

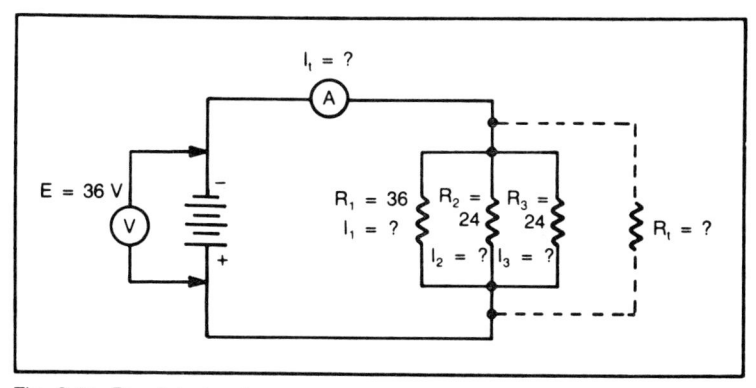

Fig. 3-14. Parallel circuit.

offers 6 Ω of resistance. It is evident then that the total current of 8 A will divide equally through both branches, join again at point B, and return to the positive terminal of the battery.

Two fundamental facts concerning the operation of a parallel circuit can now be started:

1. In a parallel Circuit, the *same* voltage is applied to each branch.
2. The total current in a parallel circuit is equal to the sum of the currents in each individual branch.

In Fig. 3-13, the total voltage of the circuit is 24 V, and the total current is 8 A. From this information, the effective resistance (R_t) may be computed by Ohm's Law:

$$R_t = \frac{E_t}{I_t}$$

$$= \frac{24 \text{ V}}{8 \text{ A}}$$

$$= 3 \text{ Ω}$$

Note that the value of R_t is *less* than that of either resistor in the circuit.

The parallel circuit in Fig. 3-14 illustrates these characteristics even further. The known values are voltage and resistances R_1, R_2, and R_3. The unknown values are current through each resistor, total circuit current, and total circuit resistance. To find these unknown values, we must apply the known facts about the action of current and voltage in a parallel circuit and use Ohm's Law.

The important known facts about this circuit are the following:

1. The voltage across each resistor, R_1, R_2, and R_3, is 36 V.
2. The sum of the currents through R_1, R_2, and R_3 will be the total current I_t.
3. The current through these resistors can be found by Ohm's Law.

$$\text{(a)} \quad I_1 = \frac{E_1}{R_1}$$

$$= \frac{36 \text{ V}}{36 \text{ }\Omega}$$

$$= 1 \text{ A}$$

$$\text{(b)} \quad I_2 = \frac{E_2 \text{ V}}{R_2 \text{ }\Omega}$$

$$= \frac{36 \text{ V}}{24 \text{ }\Omega}$$

$$= 1.5 \text{ A}$$

$$\text{(c)} \quad I_3 = \frac{E_3 \text{ V}}{R_3 \text{ }\Omega}$$

$$= \frac{36 \text{ V}}{24 \text{ }\Omega}$$

$$\begin{aligned} I_3 &= 1.5 \text{ A} \\ I_t &= I_1 + I_2 + I_3 \\ &= 1 \text{ A} + 1.5 \text{ A} + 1.5 \text{ A} \\ &= 4 \text{ A} \end{aligned}$$

The total resistance R_t can now be found by using Ohm's Law.

$$R_t = \frac{R_t}{I_t}$$

$$= \frac{36 \text{ V}}{4 \text{ A}}$$

$$= 9 \, \Omega$$

Once again, note that R_t in a parallel circuit is less than the individual resistance of any resistor in the circuit.

Laws for Parallel Circuits

Our observations about parallel circuits can be summarized in three fundamental laws that apply to any parallel circuit:

1. The same voltage is applied to each individual branch.
2. The total current is equal to the sum of the currents in the individual branches.
3. The total resistance is equal to the applied voltage divided by the total current, and this value is always less than the smallest resistance in the circuit.

Parallel Resistor Combinations

Many times you must know in advance what the total resistance of a parallel combination will be. Although Ohm's Law can be used if a known voltage and current are readily available, this approach is sometimes impractical, so you must compute total resistance by using two different formulas developed from Ohm's Law.

Reciprocal Method. Our first formula is called the *reciprocal method*, and it may be applied to any number of parallel resistances. Using the three resistances in Fig. 3-14 and the reciprocal method, we compute R_t as follows:

$$R_t = \frac{1}{\dfrac{1}{R_1} + \dfrac{1}{R_2} + \dfrac{1}{R_3}}$$

$$= \cfrac{1}{\cfrac{1}{36} + \cfrac{1}{24} + \cfrac{1}{24}}$$

$$= \cfrac{1}{\cfrac{2}{72} + \cfrac{3}{72} + \cfrac{2}{72}}$$

$$= \cfrac{1}{\cfrac{8}{72}}$$

$$= \cfrac{72}{8}$$

$$= 9\ \Omega$$

Another example is given by Fig. 3-15.

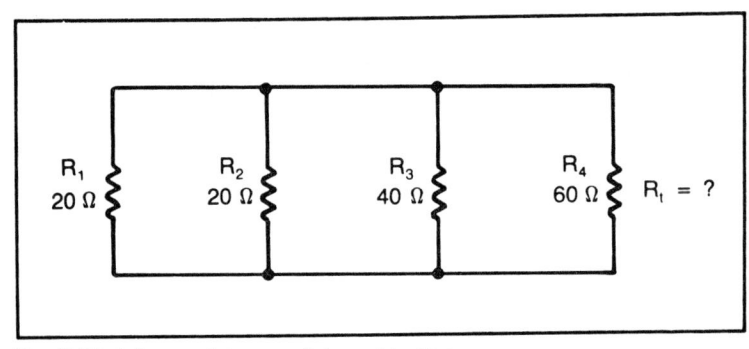

$$R_t = \cfrac{1}{\cfrac{1}{R_1} + \cfrac{1}{R_2} + \cfrac{1}{R_3} + \cfrac{1}{R_4}}$$

$$= \cfrac{1}{\cfrac{1}{20} + \cfrac{1}{20} + \cfrac{1}{40} + \cfrac{1}{60}}$$

Fig. 3-15. A four-resistor parallel combination.

$$= \frac{1}{\frac{6}{120} + \frac{6}{120} + \frac{3}{120} + \frac{2}{120}}$$

$$= \frac{1}{\frac{17}{120}}$$

$$= \frac{120}{17}$$

$$= 7.05 \; \Omega$$

These two example indicate that you can calculate the total resistance of any parallel combination of resistors without actually placing them in a circuit and measuring the voltage and current.

Like Method. This method is used when two resistors with the same value are placed in parallel. In this case the total resistance is equal to the value of one resistor divided by the total number of resistors. Figure 3-16 shows two resistors of like value in parallel. The total resistance is

$$R_t = \frac{36 \; \Omega}{2 \; \Omega}$$

$$R_t = 18 \; \Omega$$

Resistors with the same ohmic values are placed in parallel and connected in an electrical circuit, the current will divide equally

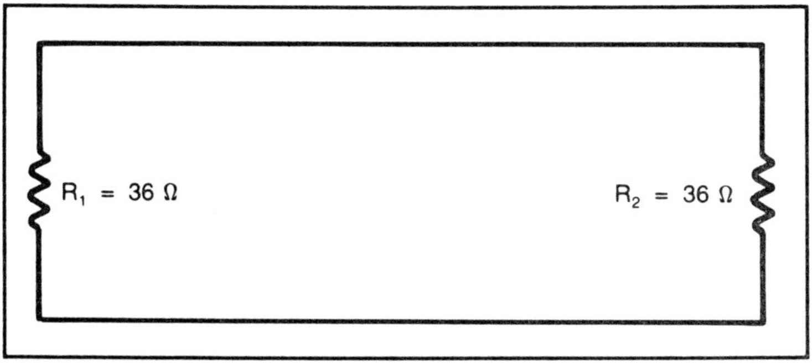

Fig. 3-16. Two equal resistors in parallel.

in the individual branches. If three like values are placed in parallel, the current in each individual branch will be one-third of the total current. If four like values are placed in parallel, as shown in Fig. 3-17, the total resistance will be one-fourth that of a single resistor.

$$R_t = \frac{16\ \Omega}{4\ \Omega}$$

$$R_t = 4\ \Omega$$

Product-Over-Sum Method. This method is applied to parallel combinations involving two resistors of unlike value. An example of this method is shown in Fig. 3-18.

$$R_t = \frac{R_1 \times R_2}{R_1 + R_2}$$

$$= \frac{24\ \Omega \times 8\ \Omega}{24\ \Omega + 8\ \Omega}$$

$$= \frac{192\ \Omega}{32\ \Omega}$$

$$= 6\ \Omega$$

To prove to yourself that the value of total resistance using the product-over-sum method is correct, use the circuit values given in Fig. 3-18 and Ohm's Law

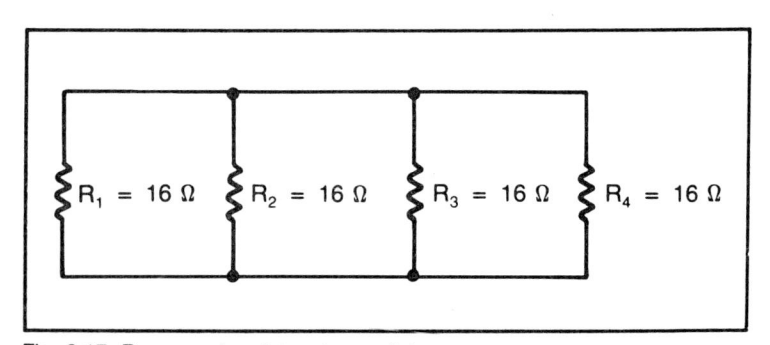

Fig. 3-17. Four equal resistors in parallel.

```
       ┌──────────────────────────────────────┐
   ─  │                                      │
  ═   E = 48 V   I₁ = 2A    R₁ = 24 Ω   I₂ = 6A   R₂ = 8Ω
  ═   │                                      │
   +  │                     R_t = 6 Ω         │
       │                                      │
       └────── I_t = 8 A ─────────────────────┘
```

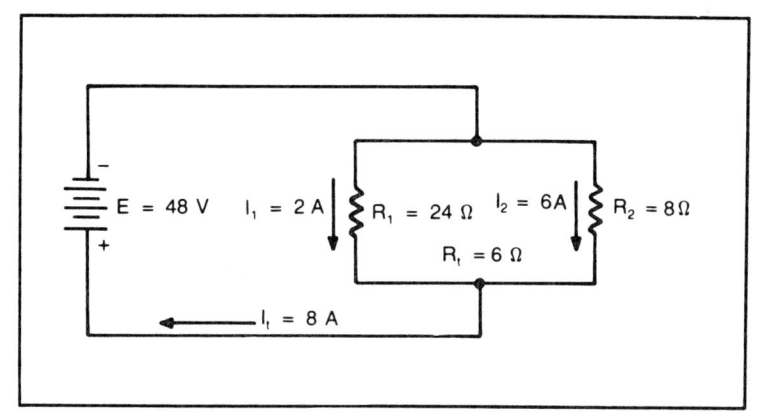

Fig. 3-18. Two resistors of unlike value in parallel.

$$R_t = \frac{E_t}{I_t}$$

To sum up, the various methods for solving for total resistances in parallel circuits are

- Ohm's Law
- The reciprocal method for any number of resistors
- The like method for resistors of the same ohmic value
- The product-over-sum method for two resistors of unlike ohmic values

What method you use depends on what the situation calls for. As you gain experience, the easiest method will become apparent.

SERIES-PARALLEL CIRCUITS

In actual practice, many electrical circuits contain devices connected in series and parallel. Figure 3-19A shows a simple series-parallel circuit Fig. 3-19B-D shows the steps used to redraw the original circuit to come up with an equivalent circuit. Some series-parallel combinations become quite complex, and all rules and laws concerning series and parallel circuits must be brought into play.

As I said earlier, putting yourself in the shoes of current flow will be helpful in solving problems in these networks. Questions such as, "If I'm current and flowing from the negative terminal of this battery, which way am I going to go through this circuit on my journey back to the positive terminal of the battery?"

53

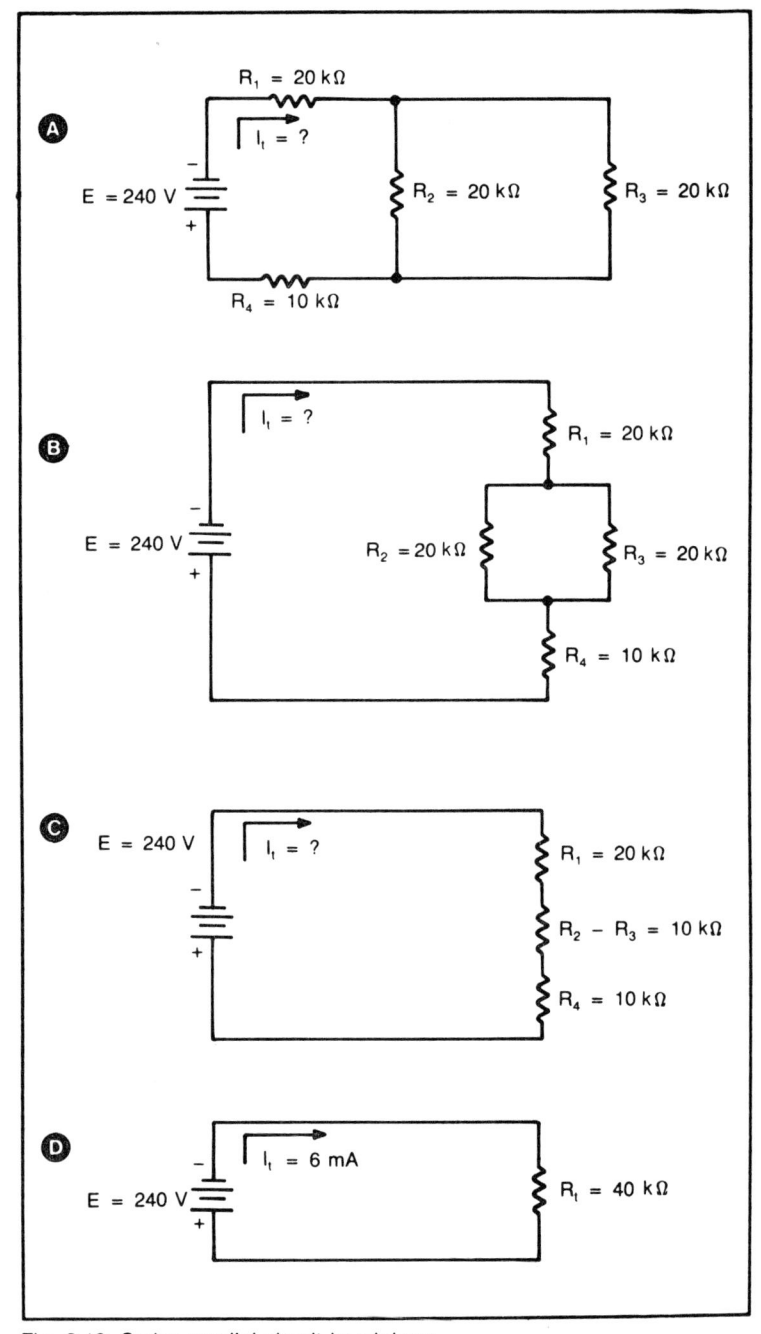

Fig. 3-19. Series-parallel circuit breakdown.

We see that Fig. 3-19A can be redrawn as shown in Fig. 3-19B. Now tracing the path that current will take, it becomes obvious that R_2 and R_3 are in parallel. The total current will divide at the junction of R_1, R_2, and R_3. This current will recombine at the junction of R_2, R_3, and R_4 and again become total current. Because the values of R_2 and R_3 are the same, we use the like method for resistors in parallel, and we find the total resistance of this combination to be 10,000 Ω.

This circuit can now be redrawn as the series circuit shown in Fig. 3-19C using the rules and laws of series circuits, we find the total resistance to be 40,000 Ω. Figure 3-19D shows the simplified circuit. We can now solve the total circuit current by using Ohm's Law:

$$I_t = \frac{E_t}{R_t}$$

$$= \frac{240 \text{ V}}{40,000 \text{ Ω}}$$

$$= 0.006 \text{ A}$$

$$= 6 \text{ mA}$$

CIRCUIT ANALYSIS

In solving complex resistive networks, you must use a logical approach. First examine the circuit; then put yourself in the current flow and determine just what path or paths you may follow

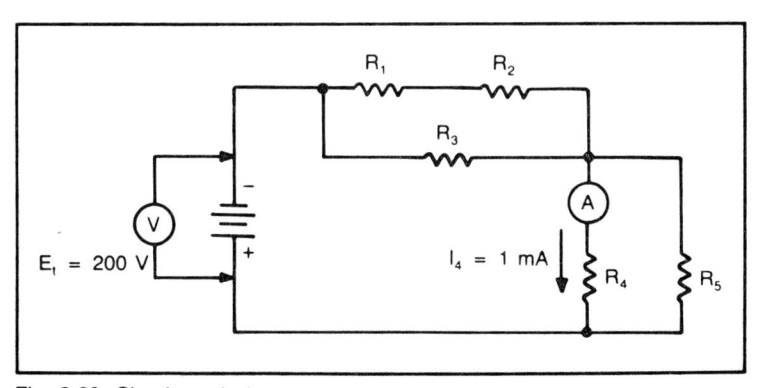

Fig. 3-20. Circuit analysis.

through the circuit in returning to the source. This first step will convert a maze into a grouping of orderly series- or parallel-connected devices.

Next, you can find the total resistance or additional values of voltage, current, or resistance to give a better overall picture of the complete circuit. In many cases only a limited number of values are known. An example is shown in Fig. 3-20. The voltmeter shows the battery voltage to be 200 V. The current through the branch containing R_4 is 1 mA. The values of R_1, R_2, and R_3 cannot be determined, but it is known that the value of R_4 is the same as the value of R_5. Find the total circuit resistance.

Step 1

Tracing the flow of current through the circuit by starting at the negative terminal of the battery, we find that the total circuit current will flow through the parallel combination of R_4 and R_5.

Step 2

Because the current flowing through the branch containing R_4 is 1 mA and the resistance value of R_4 is the same as R_5, the same amount of current will flow through R_5.

Step 3

Because the total circuit current passes through the parallel combination of R_4 and R_5, the total circuit current is

$$\begin{aligned} I_t &= I_4 + I_5 \\ &= 1 \text{ mA} + 1 \text{ mA} \\ &= 2 \text{ mA} \end{aligned}$$

Step 4

We can now use Ohm's Law to find the total circuit resistance R_t.

$$\begin{aligned} R_t &= \frac{E_t}{I_t} \\ &= \frac{200 \text{ V}}{0.002 \text{ A}} \\ &= 100{,}000 \text{ }\Omega \end{aligned}$$

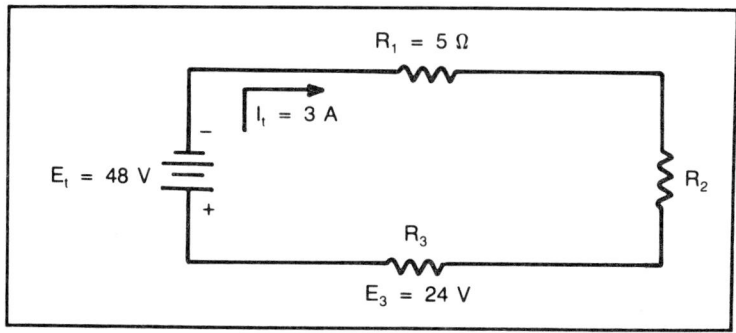

Fig. 3-21.

QUIZ

1. Give the formula for
   (a) Ohm's Law for current
   (b) Ohm's Law for voltage
   (c) Ohm's Law for resistance
2. If the voltage applied to an electrical circuit is doubled and the resistance is kept constant, what happens to the current?
3. If three resistors of 125, 95, and 65 Ω are connected in series, what is the total resistance?
4. If two resistors of 150 Ω each are connected in parallel, what is the total resistance?
5. What voltage is required to produce a current of 4 through a resistor of 24 Ω?
6. If an electric iron dissipates 1150 watts of power in the form of heat and is being operated from 115 V, what is the current that is being drawn?
7. In Fig. 3-21 find R_t, R_3, R_2, E_1, and E_2.
8. In Fig. 3-22 find R_t, R_3, E_1, and I_1.
9. In Fig. 3-23 find R_t.

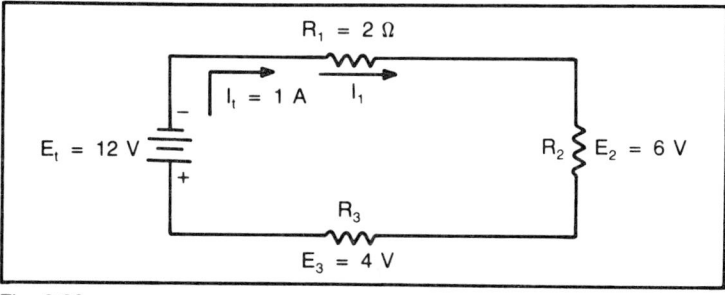

Fig. 3-22.

57

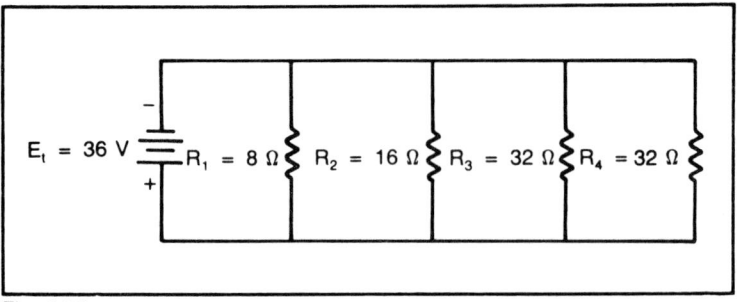

Fig. 3-23.

ANSWERS TO POWERS OF 10 EXERCISES ON PAGE 32.

(a)   (b)

5.5×10^5 1700
7.65×10^5 0.0086
8.65×10^2 880
8.390×10^3 45,000,000
4.51×10^1 2,800,000
1.875×10^6 0.00079
5.55×10^4 400,000
1.82×10^2 3400
2.96×10^5 4110
1.59×10^7 0.00000971
8.51×10^3 0.00096
1.16×10^4 0.002
5.5×10^3
3.29×10^6
2.37×10^1
1.0×10^{-4}
3×10^{-7}
1.045×10^{-3}
3.8×10^{-3}
4.2×10^{-4}

Chapter 4

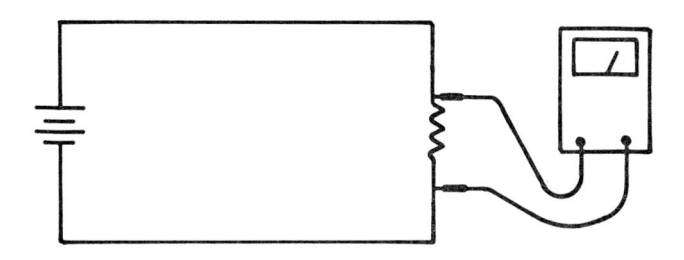

Magnetism

ONE OF THE MOST INTERESTING OF ALL ELECTRICAL PHENOMENA is magnetism. We have all seen this phenomenon in action. A material with the ability to attract iron or iron alloys is called a *magnet*. Materials attracted by magnets are said to be *magnetic*. Examples are iron, steel, nickel, cobalt, and alloys of these elements. Materials not attracted by magnets, such as wood, paper, glass, copper, or tin, are said to be *nonmagnetic*. A magnet can attract a magnetic material by actual contact or from a distance, and even through nonmagnetic material.

Magnetism was first observed with the discovery of a stone near the city of Magnesia in Asia Minor. The stone was called *magnetite* after the city. The name "magnet" was finally derived from the name of this stone.

We now know that magnetite is an iron ore possessing magnetic qualities when found in its natural habitat. Magnetite or lodestone is said to be a natural magnet. Natural magnets no longer have any useful purpose because we can produce far better magnets by artificial means.

RETENTIVITY

Different types of magnetic materials have different abilities to become magnetized and to retain this condition once they are magnetized. The ability of any magnetic material to retain its magnetism is called the *retentivity* of the material. Steel and iron alloys have a high retentivity, and materials such as soft iron and nickel have a low retentivity. The magnetism remaining in a material once the material is magnetized is called *residual magnetism*. The amount of residual magnetism remaining in a material depends upon the three factors:

- The time that the material is in contact with the inducing magnet
- The retentivity of the material
- The strength of the inducing magnetic

The most common type of magnet used is the *permanent magnet* (PM). A permanent magnet has a high retentivity; that is, it retains its magnetism for a considerable time after it is separated from the magnetizing force. If steel rather than soft iron filings are used (see Fig. 4-1), the same results are obtained, but the steel filings will retain their magnetism for a longer time after separation from the bar magnet.

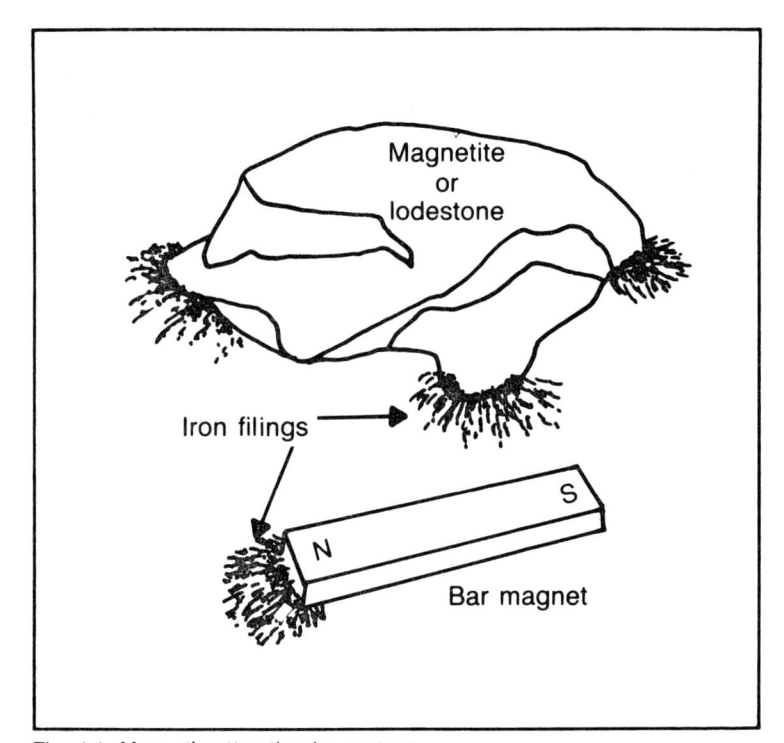

Fig. 4-1. Magnetic attraction by contact.

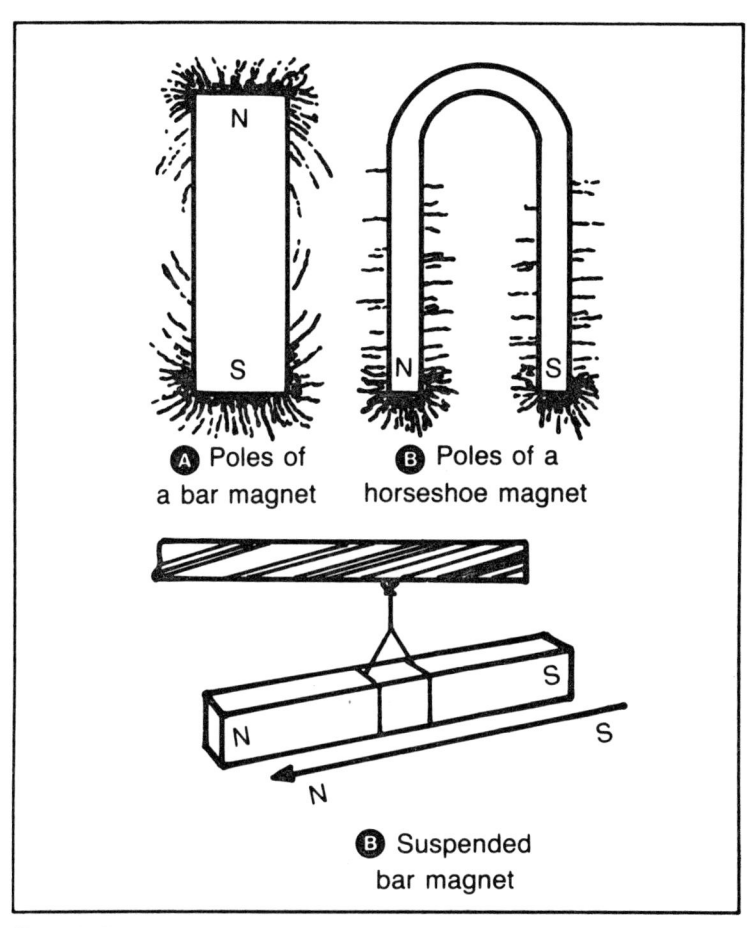

Fig. 4-2. Poles of a magnet.

MAGNETIC POLES

The overall effects of magnetism are not distributed uniformly over the entire surface of a magnet. The effects are strongest at the ends and weakest in the middle of any magnet. The regions where the magnetic effects are greatest are called the *poles* of a magnet. Figure 4-2 illustrates the poles of a bar magnet. When iron filings are sprinkled over the entire area of the magnet, those near the ends are attracted to the magnet and form bunches there. Few filings near the center are attracted by the magnet.

Figure 4-2B is a bar magnet that is purposely shaped so that the poles are closer together.

61

Figure 4-2C shows that when a bar magnet is suspended so that it is free to swing on a horizontal plane, the magnet will swing around and come to rest with one of its ends always pointing to the north. When this fact was first observed, it was decided to call the north-seeking end of the magnet the *north pole* and the south-seeking end the *south pole*. These assignments for the poles of a magnet are still used: permanent magnets are marked N or (+) at the north pole and S or (–) at the south pole.

THE MAGNETIC EARTH

The earth is surrounded by a magnetic field. The distribution of this magnetic field is the same as that which might be produced by a large bar magnet located in the center of the earth. Figure 4-3 illustrates this phenomenon. The magnetic axis of the earth in

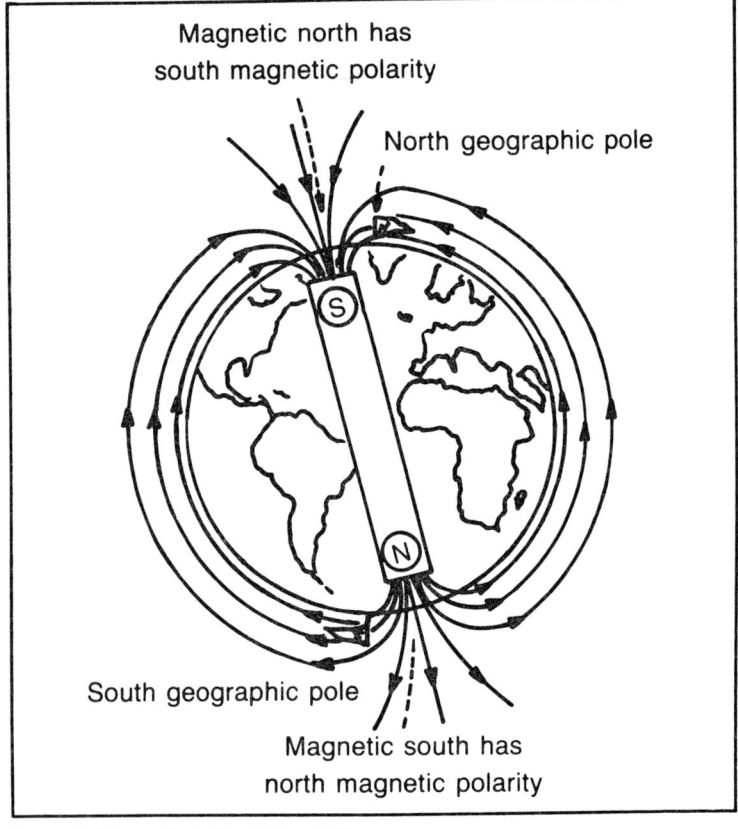

Fig. 4-3. Magnetic field of the earth.

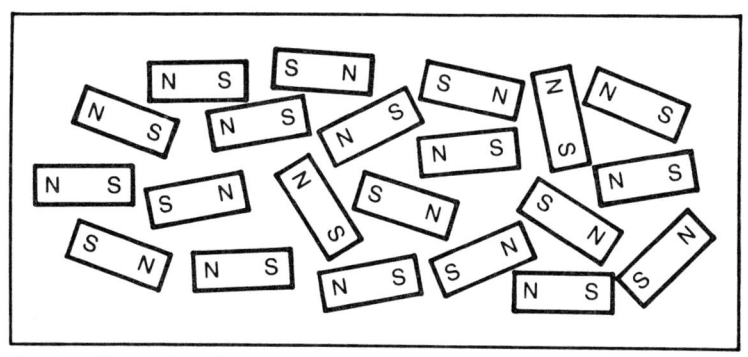

Fig. 4-4. Illustration of unmagnetized molecules.

relationship to its geographical axis is about 15 degrees. The lines of force in the earth's magnetic field flow out from the southern hemisphere and enter the surface of the earth in the northern hemisphere.

THEORY OF MAGNETISM

Many theories have been developed about magnetism, but the most popular is Weber's theory. Weber's theory is based on the assumption that all of the molecules in a magnetic material are tiny individual magnets.

When any permanent magnet is broken in half, each part becomes a magnet with its own north and south poles. If these two halves are each broken in two, four magnets now exist. Regardless of how many times these divisions are made, the parts of the original magnet are always found to be magnets themselves. If this breaking process is continued until each part is small enough to be a molecule, the molecule is still a magnet.

When a material is unmagnetized, its molecular magnets are randomly distributed (see Fig. 4-4). But when this material contacts a strong magnet, the small molecular magnets begin to line themselves up in a definite direction. As the strength of this magnetizing force is increased, more and more of the molecular magnets align in the same direction (see Fig. 4-5). When all molecules are aligned in the same direction, the material is said to be *saturated.*

LAW OF ATTRACTION AND REPULSION

When bar magnet is suspended so that it is free to swing in

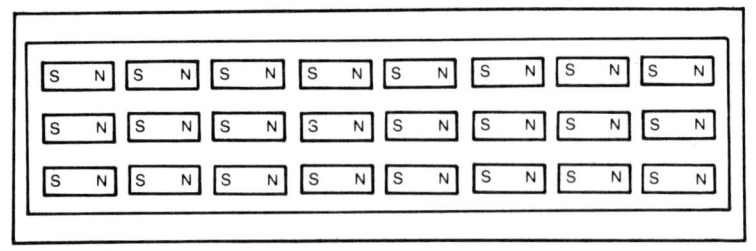

Fig. 4-5. Illustration of magnetized molecules.

a horizontal plane and the north pole of a second magnet is brought toward the north pole of the suspended magnet, the suspended magnet will be pushed away. The same thing happens if the south pole of the second magnet is brought toward the south pole of the suspended magnet. When the south pole of the second magnet is brought close to the north pole of the suspended magnet, the suspended magnet will be pulled toward (attracted) the second magnet. The same thing happens if the north pole of the second magnet is brought close to the south pole of the suspended magnet.

The above observations are used as the basis for the fundamental law of magnetism: *Like magnetic poles repel each other. Unlike magnetic poles attract each other.* Fig. 4-6 illustrates this law of attraction and repulsion.

MAGNETIC FORCE AND ITS MEASUREMENT

Because iron filings are attracted by a magnet, the magnet is said to exert a force upon the iron. The fact that this force acts even at a distance can be observed by noting the action of the magnet on iron filings, even though the magnet and filings are not in direct contact. Note that a magnet can be attracted to a piece of iron just as strongly as the iron is attracted to the magnet.

The force of attraction or repulsion between the two magnetic poles varies with the distance between them. When the poles are separated by a great distance, no visible effects are noticed. When like poles are brought close to each other, the suspended magnet is repelled (see Fig. 4-6). This force actually varies *inversely* with distance. In other words, if the distance between the two poles is halved, the force will become four times as great; if the distance between the poles is doubled, the force will become one fourth of the original amount.

Another factor involved is the strength of a pole. The force of attraction and repulsion between two magnets varies with the amount of force that their individual poles of the magnets are capable of exerting. The strength of a pole varies with its size, material composition, and degree of magnetization.

The force between two magnetic poles is directly proportional to the product of the two pole strengths and inversely proportional to the square of the distance between the two poles. Mathematically, this is

$$F = \frac{m_1 m_2}{ud^2}$$

where: F = force (in dynes) of attraction or repulsion between poles
m_1, m_2 = pole strengths (in unit poles) of the poles
d = distance (in centimeters) between poles
u = permeability of the magnetic material

Permeability is a factor that indicates the ease with which the magnetic lines of force can pass through a medium. The term

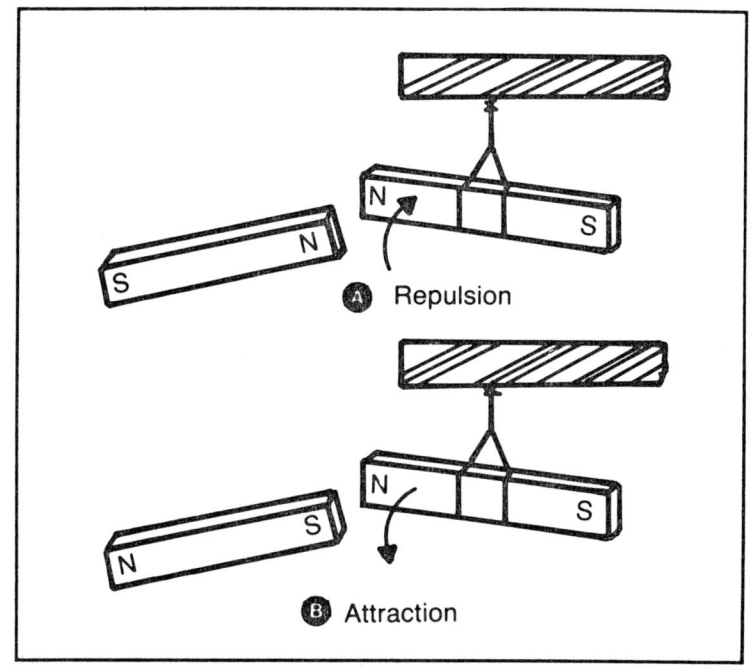

Fig. 4-6. Attraction of like and unlike magnetic poles.

"permeability," like the term "dielectric constant," is a ratio of the passage of magnetic lines of force in the given medium with respect to the passage of magnetic lines of force through air. Permeability is an important term and will be repeatedly used in this chapter.

MAGNETIC FIELD

Because a magnetic pole acts at a distance upon other poles and other magnetic materials, there must be an invisible force existing in the space surrounding a magnet. The complete magnetic field consists of an external field plus the field flowing through the material of the magnet.

A magnetic field is called a force because it has direction. A magnetic pole placed in the field moves in the direction of this force: a north pole moves in one direction, a south pole in another. To find the direction of a magnetic field, you must know the polarity of a test pole placed in the field. The *positive* direction of a magnetic field at any point is the direction in which a free north pole placed at that point would tend to move.

When a magnetic compass is placed near the south pole of a bar magnet, the compass needle will swing about and come to rest with its north-seeking pole as close as possible to the south pole of the magnet. If the compass is placed near the north pole of the bar magnet, the needle will swing and come to rest with the south pole of the needle as close to the north pole as possible. The law of attraction and repulsion between the unlike and like magnetic poles explains the action of the compass needle. See Fig. 4-7.

When the compass is placed near the center of the bar magnet, the force existing in the magnetic field causes the needle to rest in the position shown (Fig. 4-7C). In Fig. 4-7D a number of compass needles have been placed at different positions in the magnetic field of the bar magnet. Note that the needles point in many different directions. This proves that a magnetic field has a definite direction.

As I said before, you can see a representation of the intensity of this magnetic field surrounding a bar magnet by sprinkling iron filings over the area around the magnet. Many filings near the poles of the magnet are attracted to the magnet, whereas most of the other filings arrange themselves as shown in Fig. 4-8. Note that the force of the magnetic field is greatest near the poles, to which most of the filings are attracted, showing very distinct alignment.

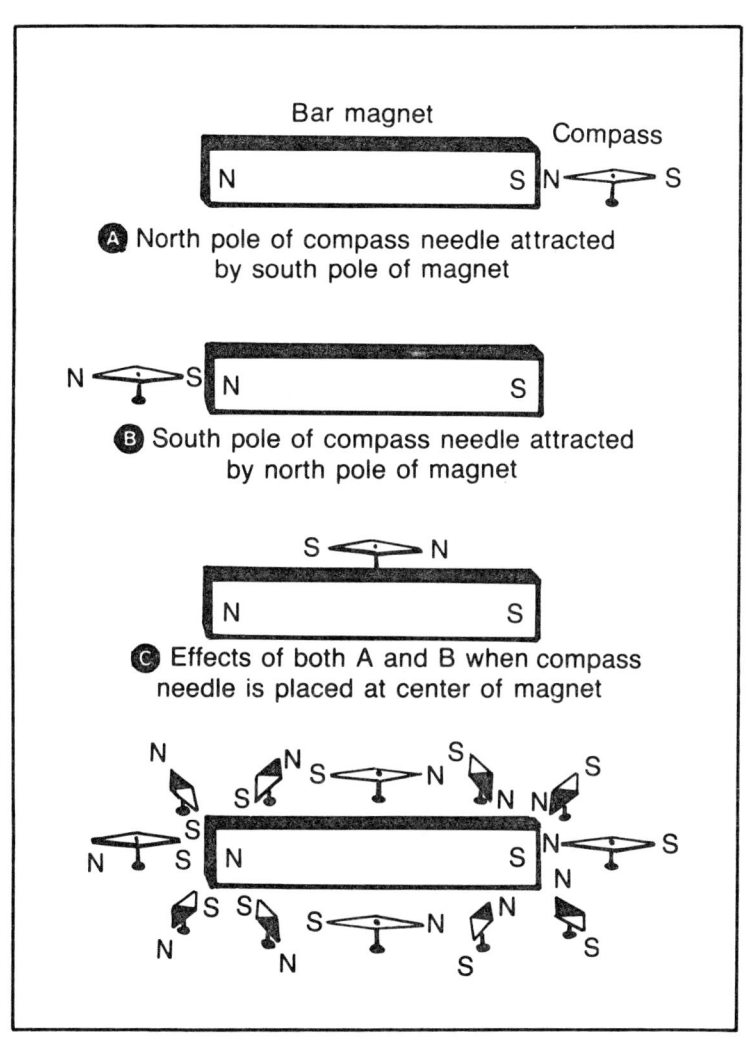

Fig. 4-7. Effects of magnetic fields on a compass needle.

As the distance from the poles increases, the field becomes less intense and the action of the force on the iron filings is less apparent. At the edge of the pattern the effect of the force on the filings is hardly noticeable. This represents the manner in which the *magnetic lines of force* in the field act on materials within the field. A study of this phenomenon lets us draw those lines of force showing the intensity and direction of the magnetic field surrounding a magnet (see Fig. 4-9).

67

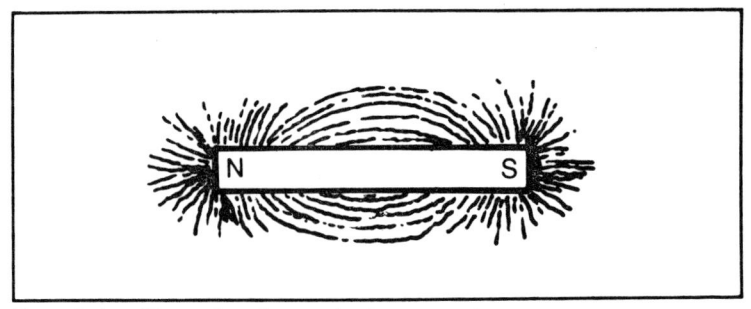

Fig. 4-8. Ion filings placed around a bar magnet.

An important fact to remember is that within the magnet itself the direction of these lines of force is from the south pole to the north pole, but outside the magnet these lines of force travel from the north pole to the south pole. A continuous loop is formed by each line of force.

The magnetic field fills the space surrounding a magnet and extends great distances from it. Theoretically, these lines of force extend in all space. Figure 4-10 shows how this magnetic field extends in all directions around the ends of a bar magnet.

Although magnetic lines of force are invisible, they have the following properties:

1. Magnetic lines of force pass through all materials, even though all materials do not have magnetic properties.
2. Magnetic lines of force that have the same direction tend to push one another apart.

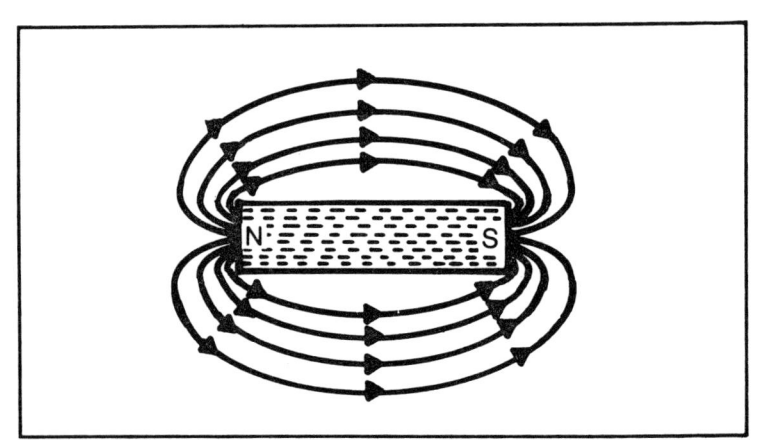

Fig. 4-9. Magnetic lines of force.

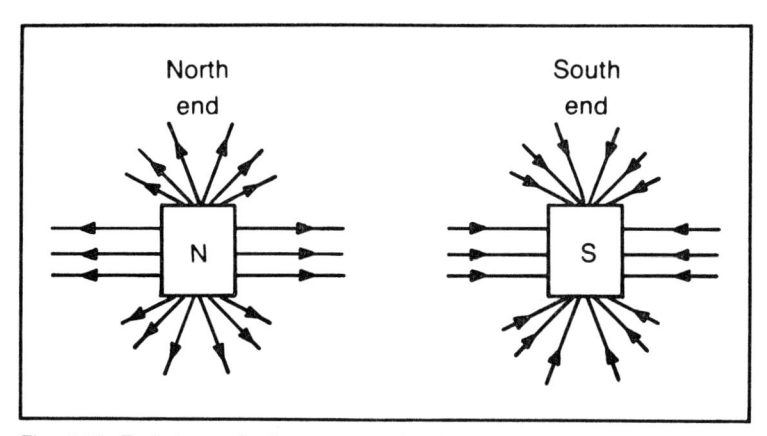

Fig. 4-10. End views of a bar magnet showing magnetic lines of force.

3. Magnetic lines of force never cross one another.
4. Magnetic lines of force are continuous and always form closed loops.
5. Magnetic lines of force have tension. When two unlike poles are brought near each other, the lines of force existing between them cause the poles (magnets) to move toward each other.

Until now we have discussed only the patterns of the magnetic fields associated with a single bar magnet. However, many patterns can be produced by using one or more magnets.

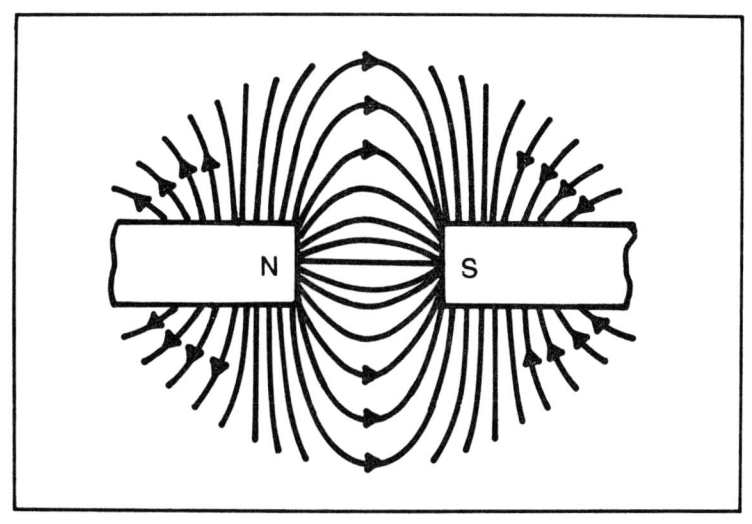

Fig. 4-11. Unlike poles exhibit forces of attraction.

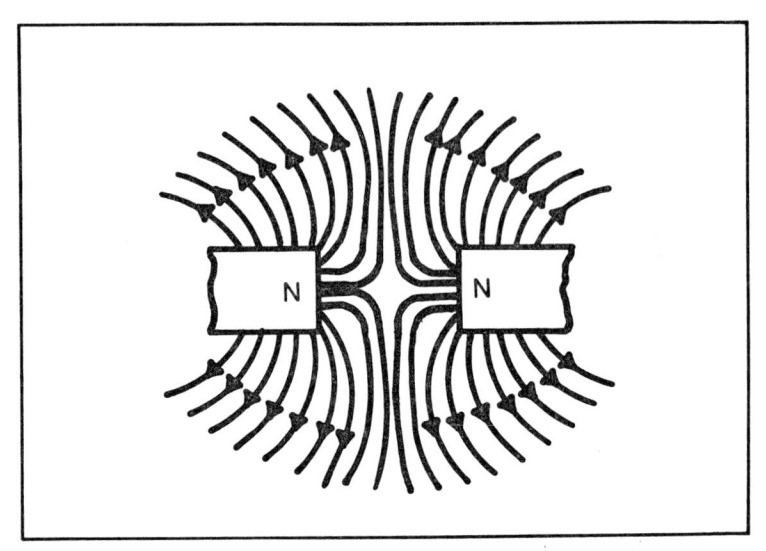

Fig. 4-12. Like poles exhibit forces of repulsion.

If the north pole of one bar magnet is brought close to the south pole of a second magnet, the pattern will look like that shown in Fig. 4-11. Note the similarity between this pattern and the pattern of the magnetic field around a single bar magnet. Figure 4-12 illustrates the manner in which these magnetic lines of force act to produce a field around two like poles. In this case the magnetic field contains forces of repulsion. Halfway between these like poles there is a magnetic field of zero intensity. This is due to the fact that the magnetic force produced by one pole is equal and opposite to the magnetic force produced by the other pole.

Figures 4-13 and 4-14 illustrate the results of placing a third magnetic pole into the fields of unlike and like magnetic poles.

Figure 4-15 shows the magnetic field associated with a horseshoe magnet. This pattern is different from the pattern of the bar magnet because of the physical position of the poles with respect to the complete magnet and to each other.

SHAPES AND USES OF MAGNETS

Magnets and their corresponding magnetic fields are very important in electrical and electronic equipment. Because of their many uses, temporary and permanent magnets come in many sizes and shapes. Bar magnets, horseshoe magnets, and ring magnets are the three primary classifications.

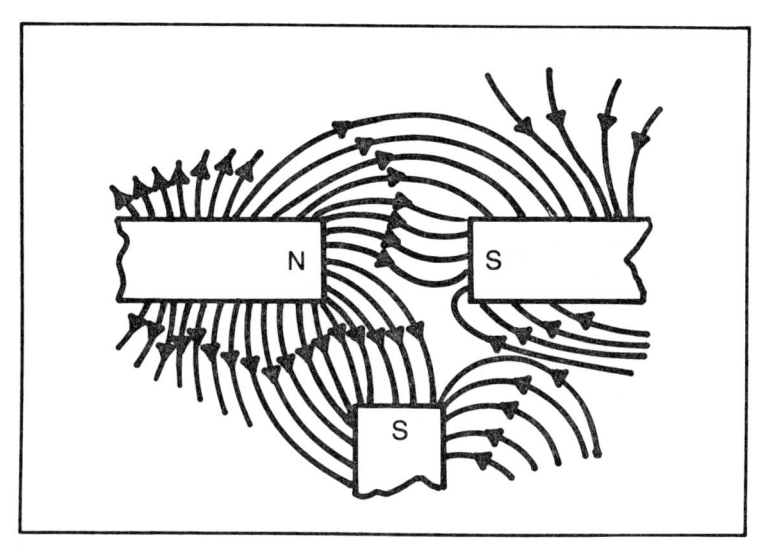

Fig. 4-13. Magnetic pattern of unlike poles with a third pole introduced.

Bar Magnets

Bar magnets are commonly used for instruction and laboratory work to study the effects of magnetism. To obtain a stronger field, place two or more bar magnets together. This is called a *compound* magnet (see Fig. 4-16). Another way is to use several thin strips of material instead of one thick strip. This type of a magnet is called

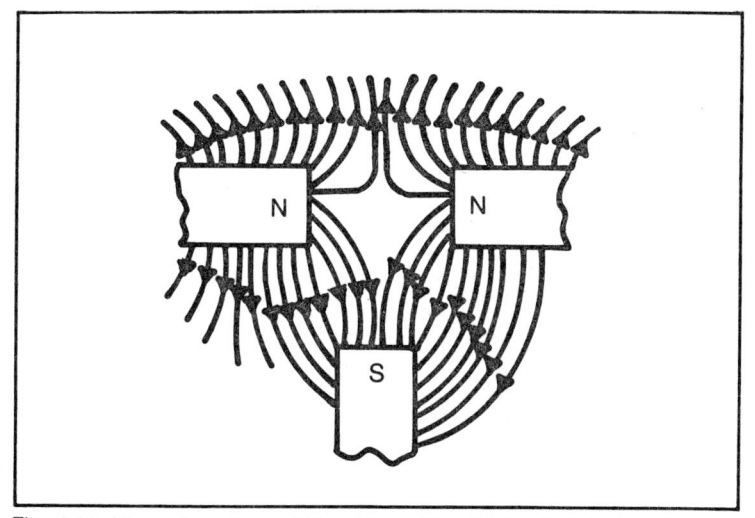

Fig. 4-14. Magnetic pattern of like poles with a third pole introduced.

71

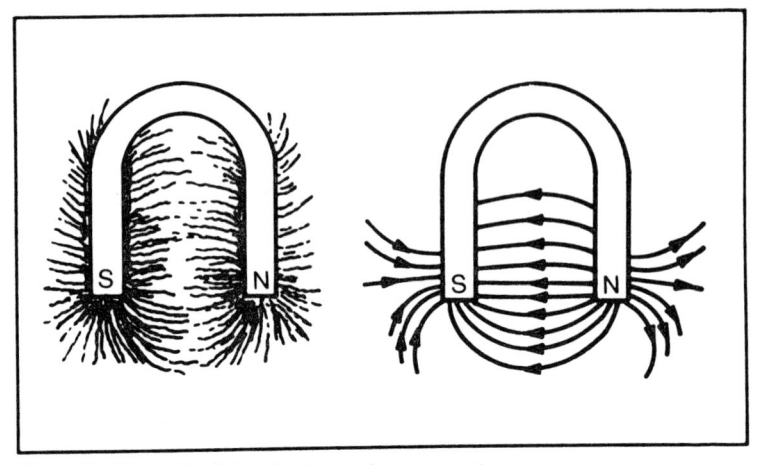

Fig. 4-15. Magnetic field of a horseshoe magnet.

a *laminated* magnet (see Fig. 4-17). For both types you must place corresponding poles together.

Horseshoe Magnets

Most electrical and electronic equipment uses horseshoe magnets because they provide a much stronger magnetic field than a bar magnet of the same material. This happens because the poles are closer together, thereby concentrating the magnetic field in a smaller space. Some shapes of horseshoe magnets are shown in Fig. 4-18.

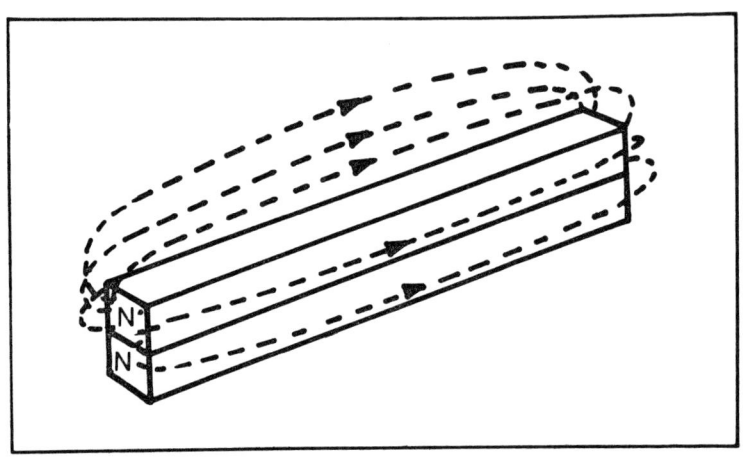

Fig. 4-16. Compound magnet.

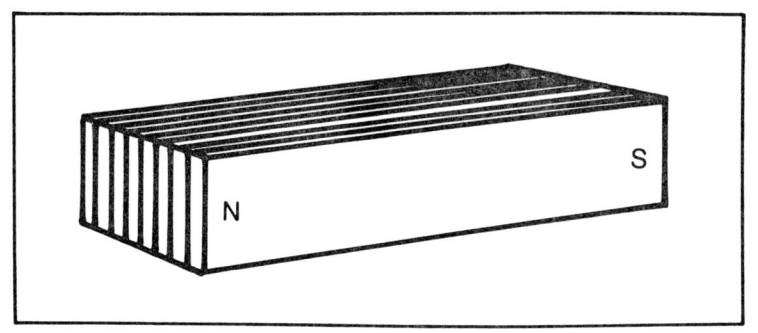

Fig. 4-17. Laminated magnet.

Ring Magnets

A ring magnet is made in a circular or square form, as shown in Fig. 4-19. Notice that the ring magnet has no magnetic poles and no outside magnetic field because the magnetic lines of force make their complete path inside the magnet. However, if a small section is cut out of the ring, as shown in Fig. 4-20, definite north and south magnetic poles are formed, and the ring magnet now becomes a form of a horseshoe magnet.

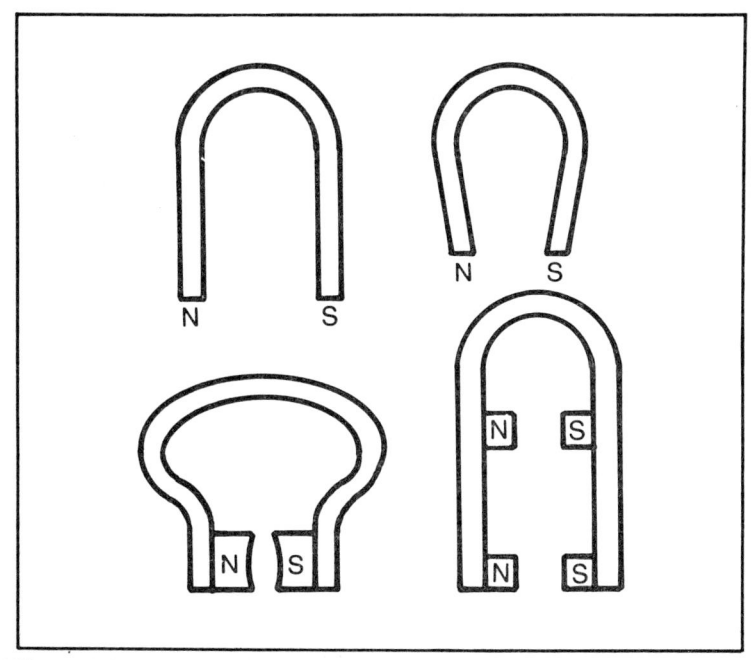

Fig. 4-18. Various shapes of horseshoe magnets.

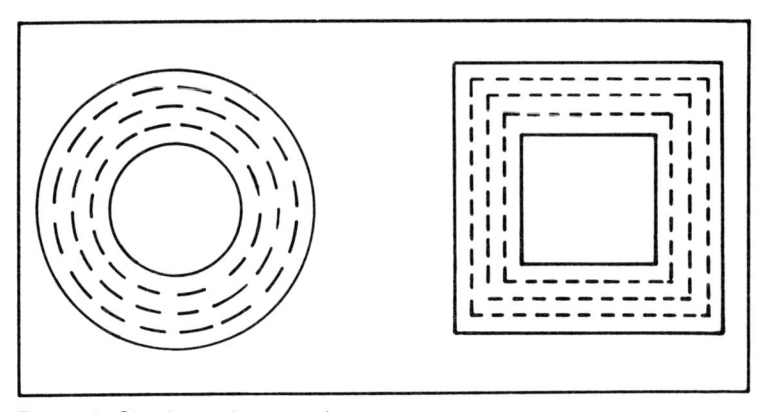

Fig. 4-19. Circular and square ring magnets.

A circular ring magnet can be used to protect sensitive instruments such as meters and watches from external magnetic forces. See Fig. 4-21. The magnetic field between the north and south poles of a magnet is distorted by the ring magnet, so that no lines of force pass through the object being shielded. The ring magnet, a temporary magnet of soft iron, provides a magnetic path for the lines of force and diverts them from the shielded object.

Although there is no known insulator of magnetic lines of force, shielding an object by placing it inside a ring magnet effectively insulates the object from the harmful magnetic field.

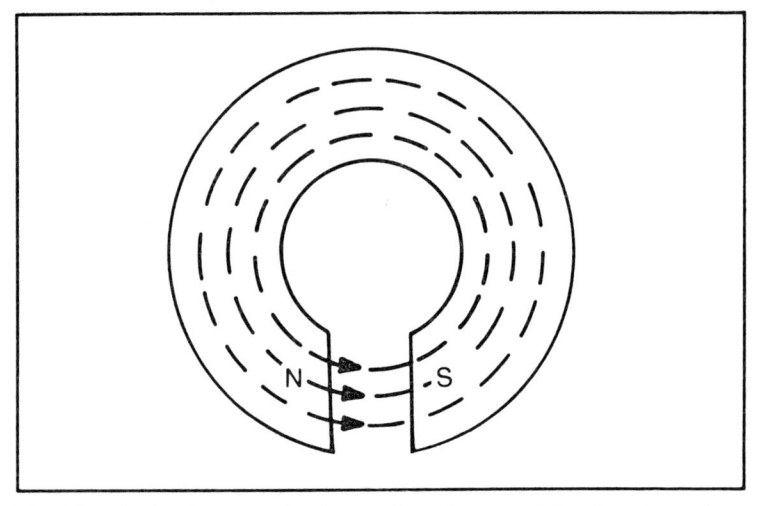

Fig. 4-20. Circular ring magnet with a small section cut out, forming a horseshoe magnet.

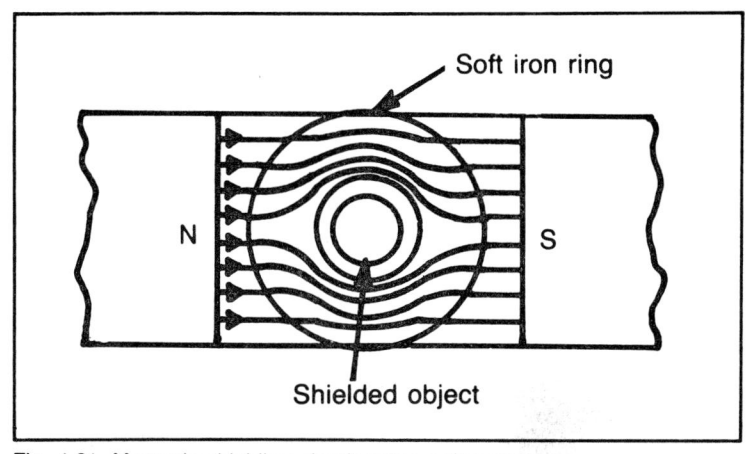

Fig. 4-21. Magnetic shielding circuit using a ring magnet.

CARE OF MAGNETS

If handled carelessly and stored improperly, permanent magnets can lose their magnetism and become useless.

A piece of steel or other magnetic material that has been magnetized may be demagnetized by an external force disturbing the aligned positions of the molecules and causing them to return to their original random positions. Heating a magnet expands the metal and allows the same molecules to return to their original position. Therefore, *never* drop a magnet or place one in a high-temperature area.

Special precautions should be taken when storing magnets. When bar magnets are not being used, they should always be stored in pairs so their adjacent ends are of opposite polarity (see Fig. 4-22).

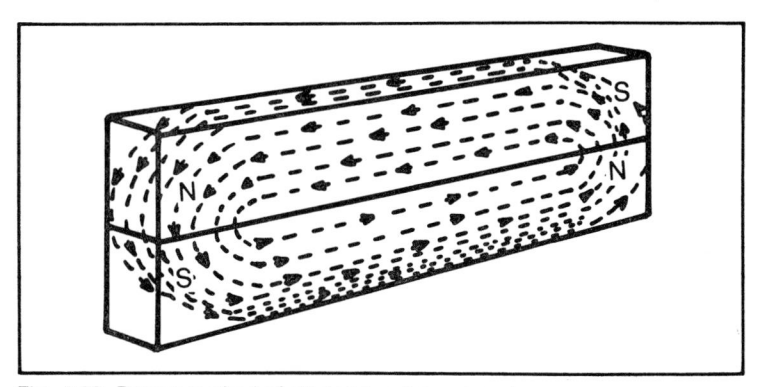

Fig. 4-22. Proper method of storing two bar magnets.

When horseshoe magnets are not being used, they should be stored with a soft iron bar called a *keeper*. The keeper should be placed across the poles to prevent flux-line leakage (see Fig. 4-23). Note that the lines of force pass from the north pole to the south pole of the magnet through the permeable path of the soft iron keeper, keeping the flux leakage to a minimum.

No special precautions are required to store ring magnets because they have no poles or outside magnetic field. These magnets are, however, also subject to destruction by jarring and high temperatures.

ELECTROMAGNETISM AND MAGNETIC CIRCUITS

Until the year 1820 the study of magnetism was considered to be distinct from the study of electricity. Then Hans Christian Oersted observed that there was a direct relationship between magnetic force and electric force. He noticed that placing a compass needle in parallel to a current-carrying conductor caused the compass needle to turn and come to rest in a position at right angles to the conductor. Continuing his experiments, Oersted found that if the current was reversed in the conductor the needle would swing at right angles in the opposite direction. Through these experiments, Oersted proved that a wire carrying an electrical cur-

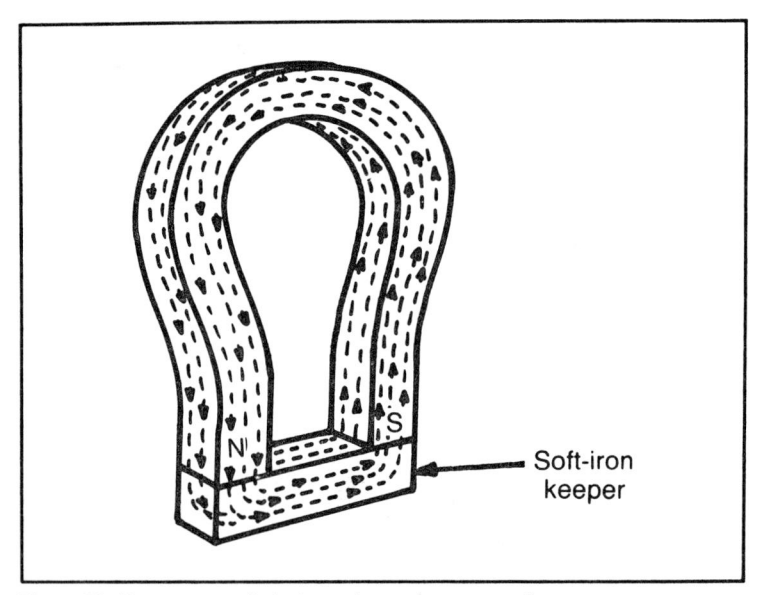

Fig. 4-23. Proper way of storing a horseshoe magnet.

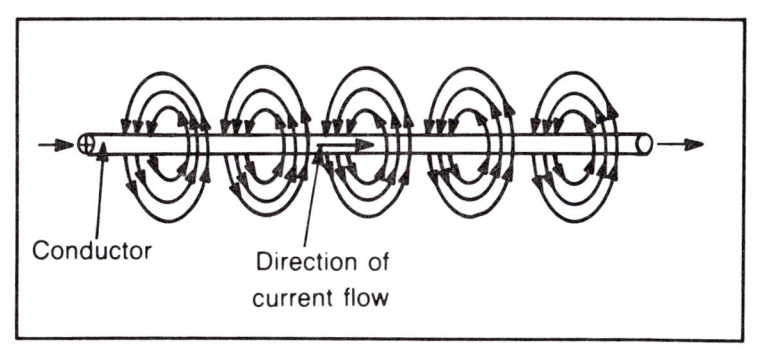

Fig. 4-24. Magnetic field surrounding a current-carrying conductor.

rent has a field of force around it that acts on a compass needle in the same way that a field of force around a permanent magnet does.

Conductor's Magnetic Field

When a conductor has a current flowing through it, a magnetic field exists about the conductor (see Fig. 4-24), which is always at right angles to the current that produces it. Because a magnetic field has both intensity and direction, the lines of force are closely concentrated near the conductor and gradually become less concentrated as the distance from the conductor increases.

If the current were increased, the compass needle would again be affected and indicate the direction of the magnetic field. This can be expressed mathematically as

$$H = \frac{2I}{d^2}$$

where: H = magnetic field strength
I = current through a conductor
d = perpendicular distance from conductor to a point where magnetic field is to be evaluated

Left-Hand Rule

After Oersted's discovery, a simple rule, known as the *left-hand rule*, was established for determining the direction of the magnetic field about a current-carrying conductor: Grasp the conductor in the left hand with the thumb pointing in the direction of the current flow, and the direction of the fingers will show the direction of the magnetic lines of force. Figure 4-25 shows how to use the left-hand rule.

Magnetic Field around a Coil of Wire

When many loops or turns of wire are wound to form a coil, it may be called a helix or a solenoid. The magnetic field surrounding a solenoid is shown in Fig. 4-26. The lines of force of this magnetic field produce a magnetic field similar to that of a bar magnet with a north pole at one end and a south pole at the other. As these loops are pushed closer together (Fig. 4-27), many more lines of force will encircle the entire solenoid. Thus, the magnetic field in the coil and at the poles is much stronger than the field of the solenoid shown in Fig. 4-27 because most of the lines of force that previously had encircled the individual loops will now encircle the entire solenoid. While the current is flowing, the solenoid has all the properties of a permanent magnet. The left-hand rule can be used to determine the direction of the magnetic field surrounding a coil in the following way. Grasp the coil with the left

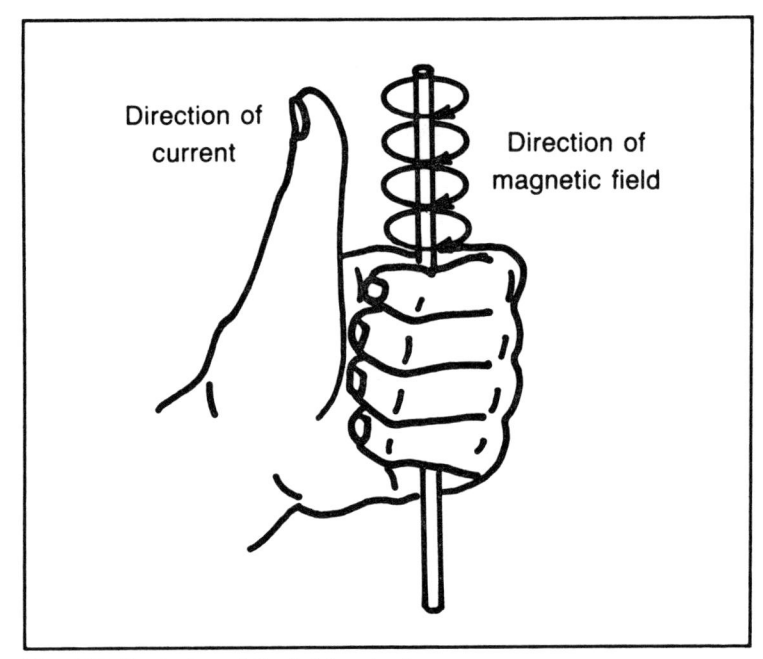

Fig. 4-25. Illustration of the left-hand rule.

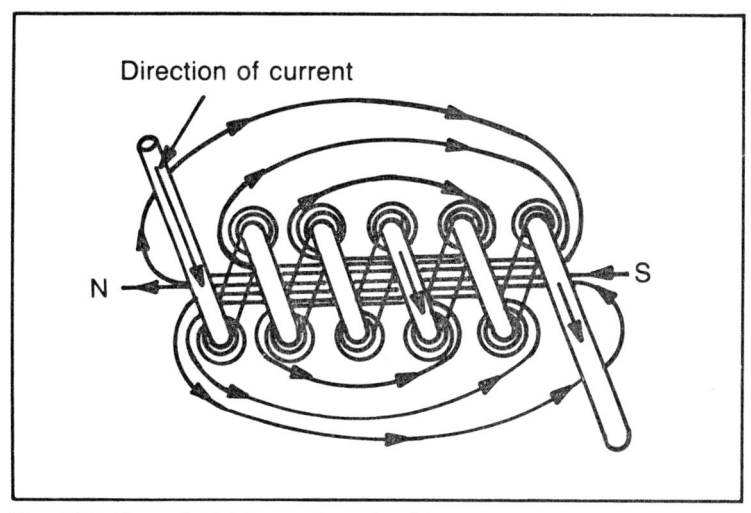

Fig. 4-26. Magnetic field around a solenoid.

hand so that the fingers point in the direction of the current flow around the coil; extend the thumb at right angles to the fingers; the thumb then points in the direction of the north pole. Figure 4-28 shows this use.

The Electromagnet

If a piece of magnetic material such as soft iron is placed within a coil, the magnetic strength of the coil is greatly increased due to the better magnetic path provided by the soft iron for the lines of force. The inside of any coil, whether air or some type of

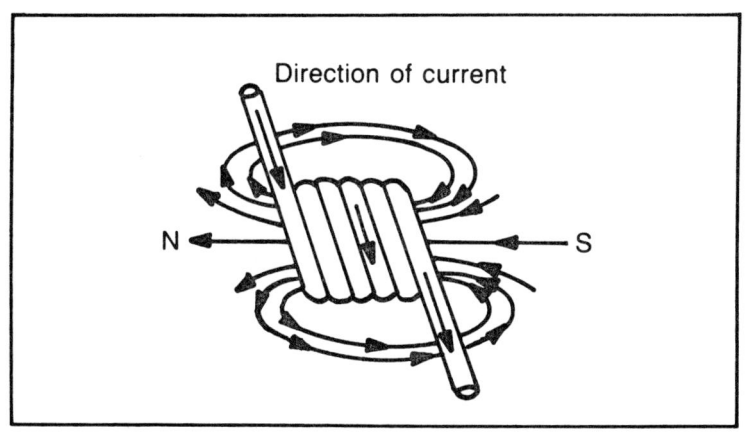

Fig. 4-27. Effect on magnetic field around a solenoid by closer loop spacing.

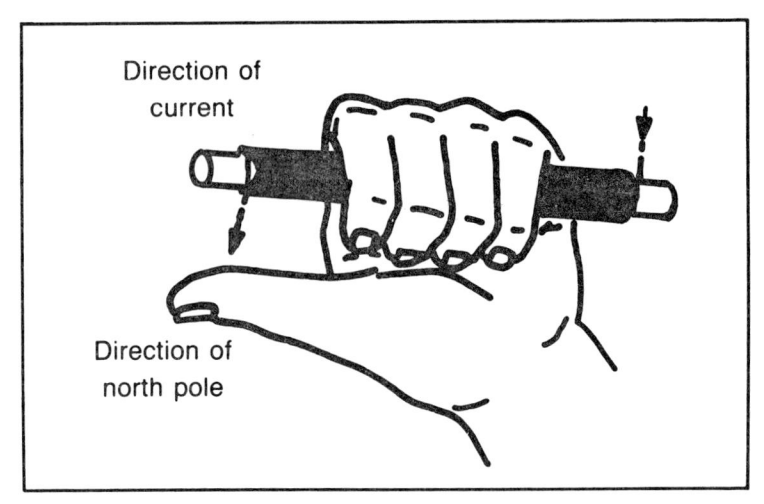

Fig. 4-28. Left-hand rule for coils.

magnetic material, is called the *core* of the magnet. If the coil is wound on a core of magnetic material, it is called an *electromagnet*. An example of a typical electromagnet is shown in Fig. 4-29. This coil may be wound with one or more layers of wire extending from one end of the core to the other. The left-hand rule for determining the direction of the magnetic field of a coil is also used to determine the direction of the magnetic field.

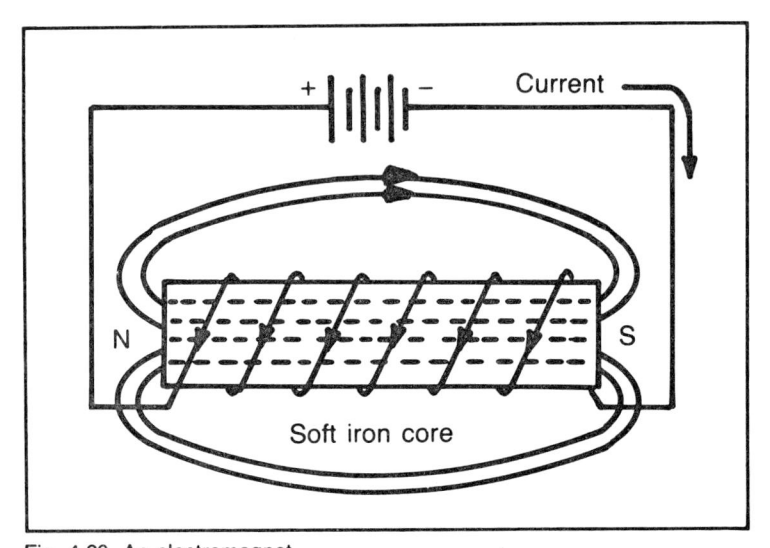

Fig. 4-29. An electromagnet.

Magnetic Circuits

The path followed by magnetic lines of force through air or through a magnetic material is called the *magnetic circuit*. The laws applying to magnetic circuits are similar to those pertaining to electrical circuits. The next section compares the properties of a magnetic circuit with those of an electric circuit.

ELECTRIC AND MAGNETIC CIRCUIT PROPERTIES COMPARED

Current flow was defined as the movement of free electrons from one atom to another. The relationship of current, voltage, and resistance for electric circuits is expressed by Ohm's Law. Comparable to the current, voltage, and resistance of electric circuits are the magnetic lines of flux, magnetomotive force, and reluctance of magnetic circuits.

The *flux*, ϕ, in a magnetic circuit is the entire quantity of magnetic lines of force in or around a magnetic circuit. The energy or force required to produce this magnetic flux through a magnetic circuit is called *magnetomotive force*, Σ (mmf). The unit of magnetic flux is the maxwell and the unit of mmf is the gilbert.

The opposition offered to the flux in a magnetic circuit is called *reluctance* (R). Reluctance is similar to resistance in electric circuits, but it is not a constant value for a particular material. Reluctance varies with flux density. *Flux density* is the number of flux lines per unit area of any material. There is no name given to the unit of measurement for reluctance. The mathematical relationships between flux, mmf, and reluctance are very similar to Ohm's Law:

$$F = \phi R$$

$$R = \frac{F}{\phi}$$

$$\phi = \frac{F}{R}$$

A comparison between current, voltage, and resistance in an electric circuit and flux, mmf, and reluctance in magnetic circuits is shown in Table 4-1.

Table 4-1. Comparison of Magnetic and Electric Circuit Properties.

Magnetic Circuits		
Property	Symbol	Formulas
Magnetomotive force	F	$F = \phi R$
Flux	ϕ	$\phi = F/R$
Reluctance	R	$R = F/\phi$
Voltage	E	$E = IR$
Current	I	$I = E/R$
Resistance	R	$R = E/I$

QUIZ

1. State the law of magnetic attraction and repulsion.
2. True or false: magnetic lines of force pass through all materials.
3. External to the magnet, lines of force flow from the _____ pole to the _____ pole.
4. Lines of force inside a magnet flow from the _____ pole to the _____ pole.
5. Does a ring magnet have any magnetic poles?
6. The unit of magnetic flux is the _____.
7. The unit of magnetomotive force is the _____.
8. The opposition offered to magnetic flux lines is called _____.
9. A bar magnet purposely shaped so its poles are closer together is called a _____.
10. When you use the left-hand rule the direction of your fingers indicates the direction of _____.

Chapter 5

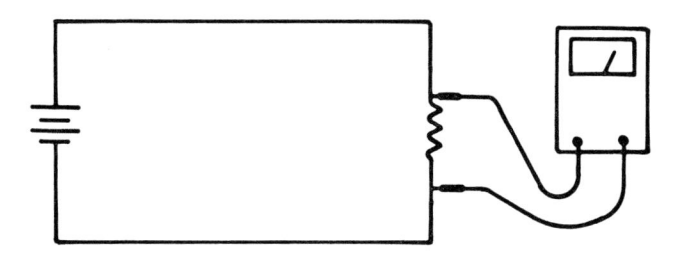

Batteries

THE FIRST USABLE SOURCE OF EMF WAS THE CHEMICAL BATtery. Two Italian scientists were deeply involved in this area. Luigi Galvani was the first to discover the process, but Allesandro Volta actually constructed the first battery.

This battery was called the *voltaic pile*. Today the voltaic pile has been refined into the highly efficient primary cell presently in massive use of such diverse applications as powering extremely small hearing aids to operating large emergency lanterns.

Another type of battery is the secondary cell, sometimes called the storage battery. The main advantage of the secondary cell is its ability to be used over a long time, provided its chemical energy is replenished at regular intervals. A typical example of a secondary cell is the 12-volt battery used in automobiles.

The history of batteries is quite interesting. Once they were the main source for operating radios and other electrical devices. Then a period of time existed when generators supplied most needs, with batteries filling in the gaps on portable and emergency equipment. New technology then raised batteries from the graveyard. This technology replaced vacuum-tube equipment, which was large and cumbersome and required heavy current drains, and batteries were called upon to power a host of electrical and electronic equipment.

THE PRIMARY CELL

A *primary cell* is classified as any type of a cell in which permanent chemical change takes place in order to produce an emf. The voltaic pile was the first primary cell (see Fig. 5-1). It is made of a rod of insulating material on which silver and zinc discs are alternately stacked, each separated from the other by cloth separators soaked in a salt solution.

Further experimentation by Volta produced an improved version, called the *voltaic cell* (see Fig. 5-2). It consists of a wire connected across copper and zinc plates in lye solution. Volta discovered that a current flow through the wire could be detected.

Electrolytes

The two dissimilar metal plates of the voltaic cell are referred to as *electrodes*. The liquid solution in which these electrodes are immersed is called the *electrolyte* and may be composed of just about anything, such as acid, alkaline, or salt dissolved in water, de-

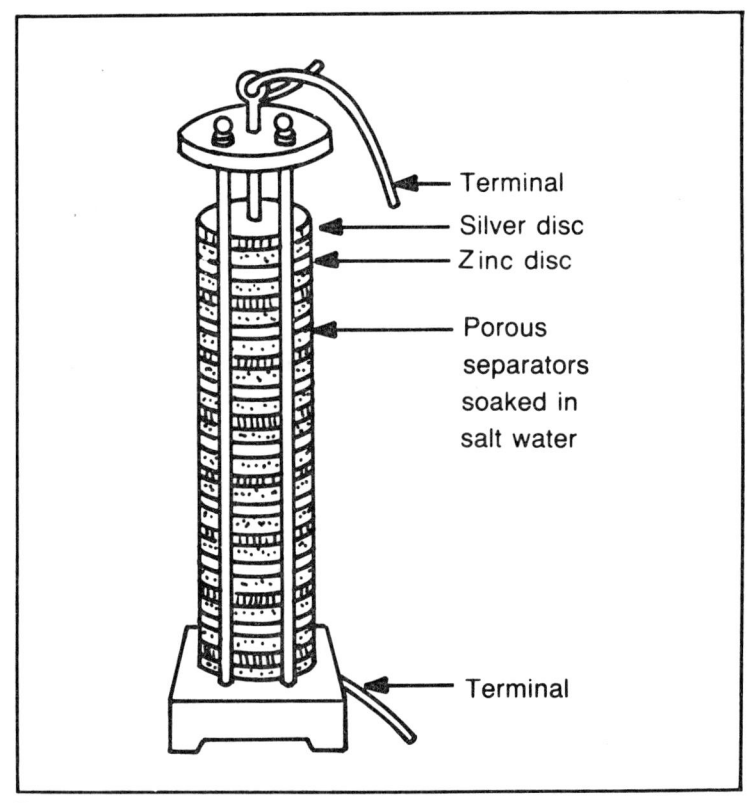

Fig. 5-1. Voltaic pile.

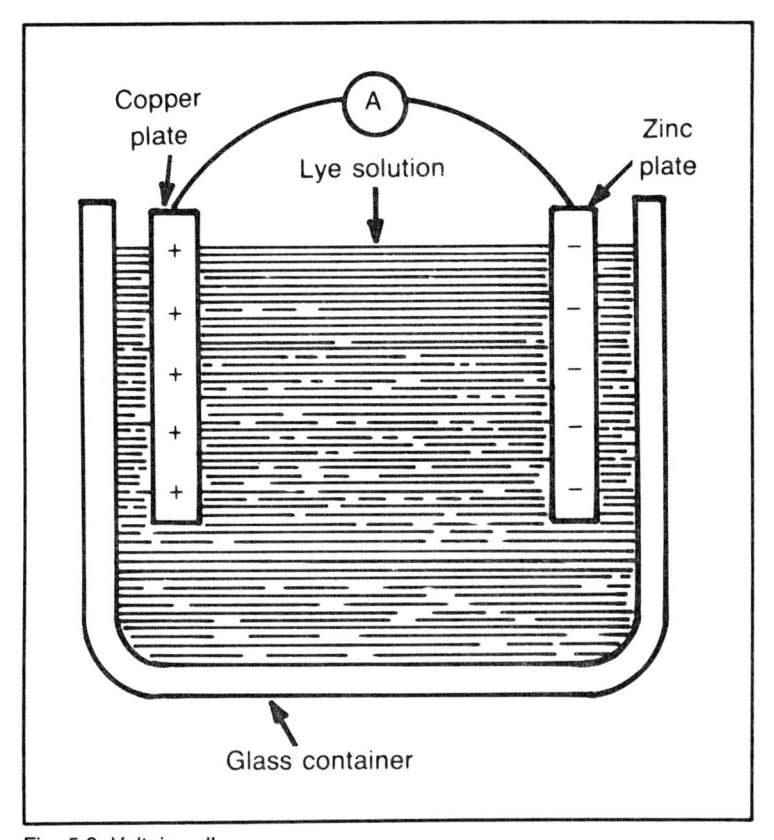

Fig. 5-2. Voltaic cell.

pending on the materials used as electrodes. While my children were growing up and participating in the various science fairs at school, a good demonstration was created that used a sensitive microammeter. Experimenters found that by inserting two electrodes in just about anything a detectable current would be present. A favorite electrolyte was citrus fruit.

The purpose of the electrolyte is to chemically act upon the two electrodes and create a difference of potential between them. When an acid or a salt is mixed with water to form an electrolyte solution, two processes take place: the acid or salt dissolves in the water; and a chemical reaction called *ionization* occurs (i.e., the atoms of the chemical partially break up into particles bearing electrical charges).

As electron flow continues, a chemical action takes place until the zinc electrode is reduced to ions and ceases to exist. Of course, current flow then ceases, and the battery or cell is said to be exhausted.

Internal Resistance

The electrolyte solution presents some opposition to the motion of the ions. For the ions to move within the cell, a part of the total voltage developed by the cell must be used. This internal voltage requirement must be subtracted from the total emf of the cell, resulting in a lowered terminal voltage when current is flowing in the external circuit. This is similar to the action of a resistance to electron flow; therefore the cell is said to have an *internal resistance*. To determine the internal resistance of a cell, we must make an open-circuit voltage measurement and a closed-circuit voltage measurement. The open-circuit voltage measurement is accomplished by placing a high-resistance voltmeter across the battery electrodes as shown in Fig. 5-3. The closed-circuit voltage is then

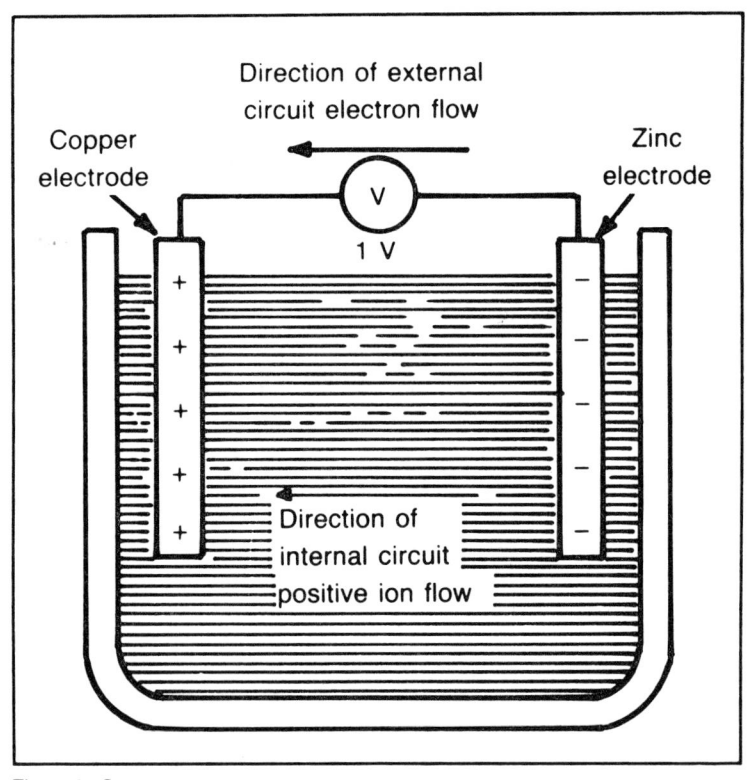

Fig. 5-3. Open-circuit measurement.

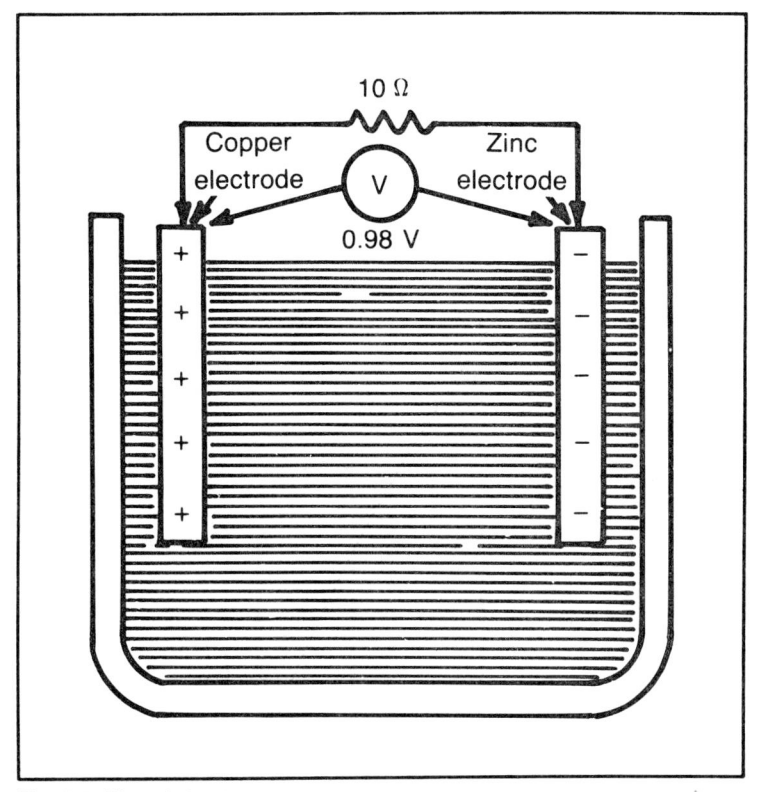

Fig. 5-4. Closed-circuit measurement.

taken with a low-resistance load placed across the electrodes as shown in Fig. 5-4. These measurements indicate a difference of 0.02 V.

BATTERY COMBINATIONS

An electrical or electronic circuit often requires a voltage or current that a single cell is incapable of supplying. To supply this need, we can group cells in various series and parallel combinations.

Series Connections

Recalling the laws of series circuits:

■ The sum of the individual voltage drops is equal to the source voltage.

■ The amount of current throughout a series circuit is the same.

These laws are also true for series-connected batteries. Figure 5-5 shows four primary cells connected so that the resultant voltage applied to the load resistor is equal to the sum of the voltages of each cell and so that the current is equal to the current capability of only one cell. Using Ohm's law, you can see that the 12-Ω load

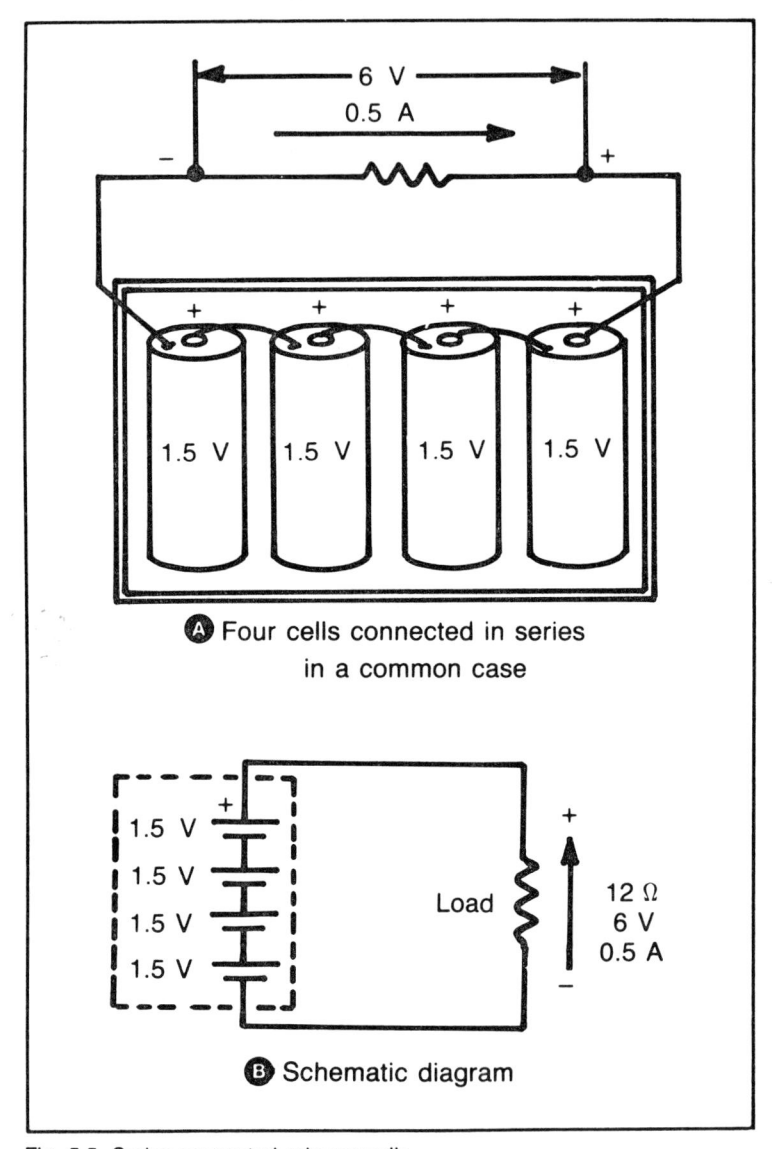

Fig. 5-5. Series-connected primary cells.

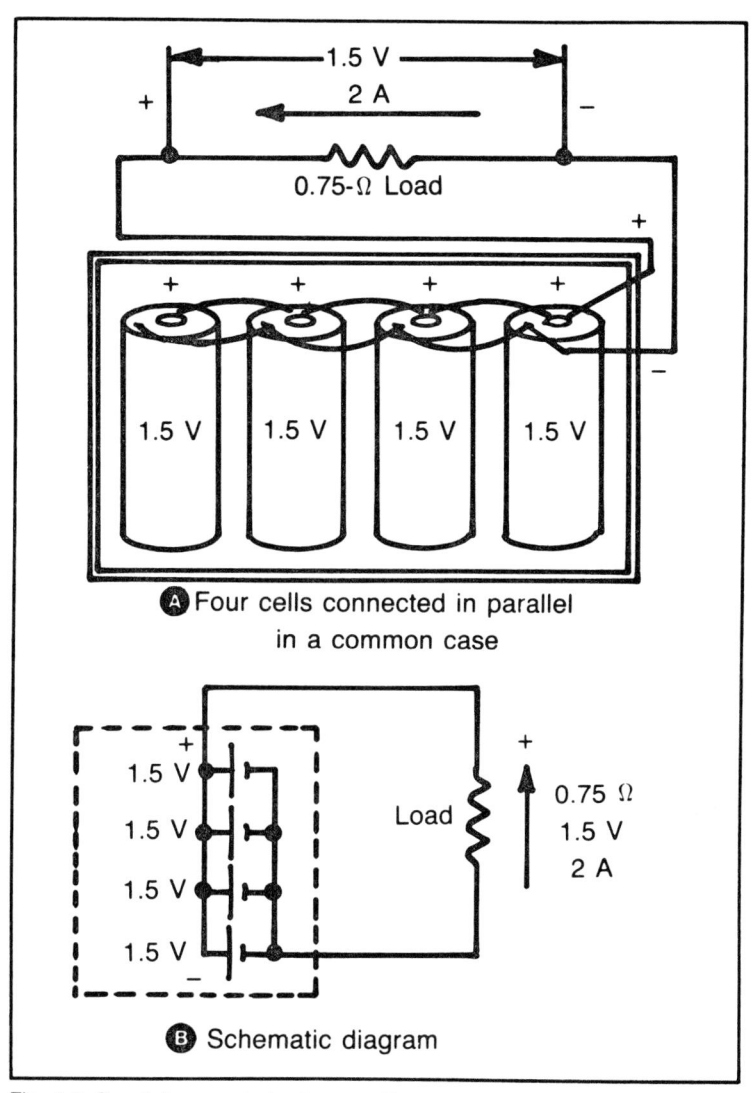

Fig. 5-6. Parallel-connected primary cells.

resistor with 6 V impressed across it must have 0.5 A of current flowing through it. If cells of unequal current capabilities are used, the current available is determined by the weakest or smallest cell.

Parallel Connections

When batteries are connected in parallel, as shown in Fig. 5-6, they conform to the laws of parallel circuits. By connecting the four 1.5-V, 0.5-A batteries in parallel, we see that the load resistor draws 0.5 A from each cell, which adds up to 2 A of current supplied from this combination.

This means that the parallel combination can provide four times the amount of current of one cell for the *life expectancy of one cell*. Because the voltage across a parallel combination is equal to the voltage across any of its branches, the output voltage of this network is only 1.5 V. For parallel arrangements of batteries, each parallel branch must have batteries or cells producing the same voltage as batteries in the other branches.

When you need a large current at a low voltage, use parallel connected batteries. Also, if you want your equipment to operate longer than one cell could provide for, use two cells in parallel to increase the current capability and the operating time.

Series-Parallel Connections

If the requirements exceed the current and voltage outputs of a single battery, you can use a series-parallel combination to furnish the desired values. For example, a circuit requiring 90 V at 4 A can be assembled by connecting two 45-V, 2-A batteries in series; this gives a battery that can supply 90 V at 2 A. This combination meets the voltage requirement, but it does not satisfy the current requirement. By connecting two pairs of these series-connected batteries in parallel, we obtain the required voltage and current. The terminal connections for this arrangement are illustrated in Fig. 5-7.

SECONDARY CELLS

The primary cell and the secondary cell produce electrical energy by transforming chemical energy. The chemical composition of the materials used in the secondary cell are altered during the transformation, but the materials themselves are not affected; therefore they are not destroyed as in the primary cell. When the chemical structures of the materials used in the secondary cell are completely changed, the production of an emf ceases.

The main advantage of the secondary cell is that the original properties of the materials can be restored, so that the battery can again furnish an emf comparable to that of a new cell.

The most common secondary cell is the lead-acid battery used to start the gasoline engines of automobiles, airplanes, and boats.

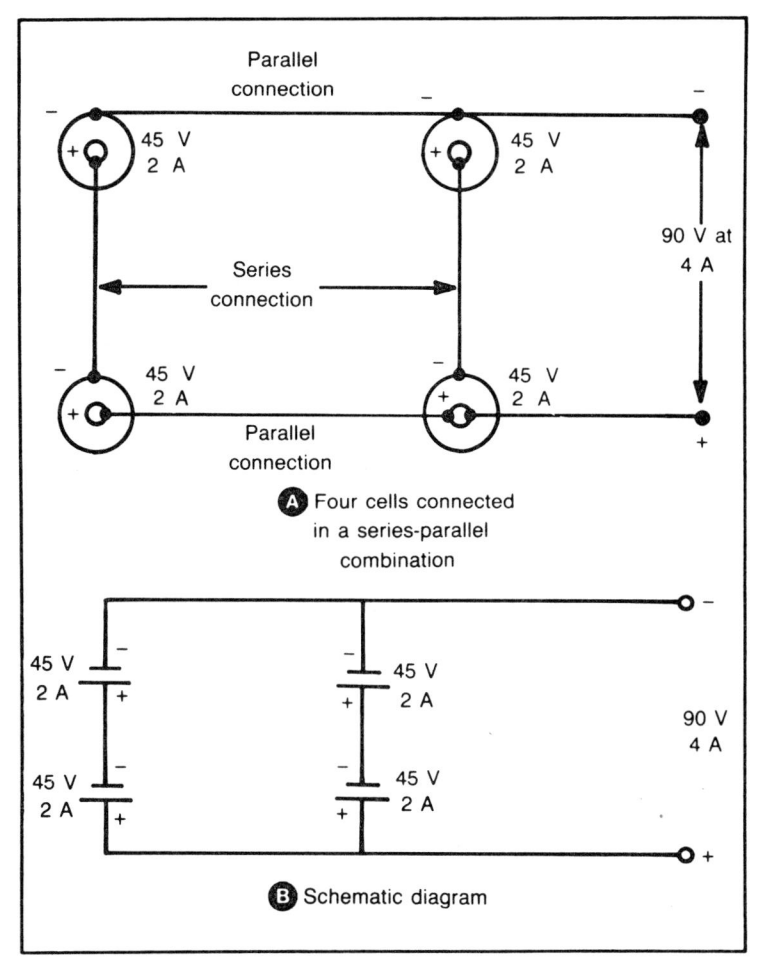

Fig. 5-7. Series-parallel-connected primary cells.

Chemical Action

A typical, charged secondary cell is shown in Fig. 5-8. The electrodes are composed of spongy lead (−) and lead peroxide (+). The electrolyte is a solution of sulfuric acid and water. The emf that this cell is capable of producing when fully charged is 2.1 V. Because a difference in potential exists, when the battery is connected to an external circuit electrons will flow from the spongy lead electrode through the circuit connection to the lead-peroxide electrode. The chemical action occurring at the lead-peroxide electrode due to the electron movement causes the lead peroxide to

91

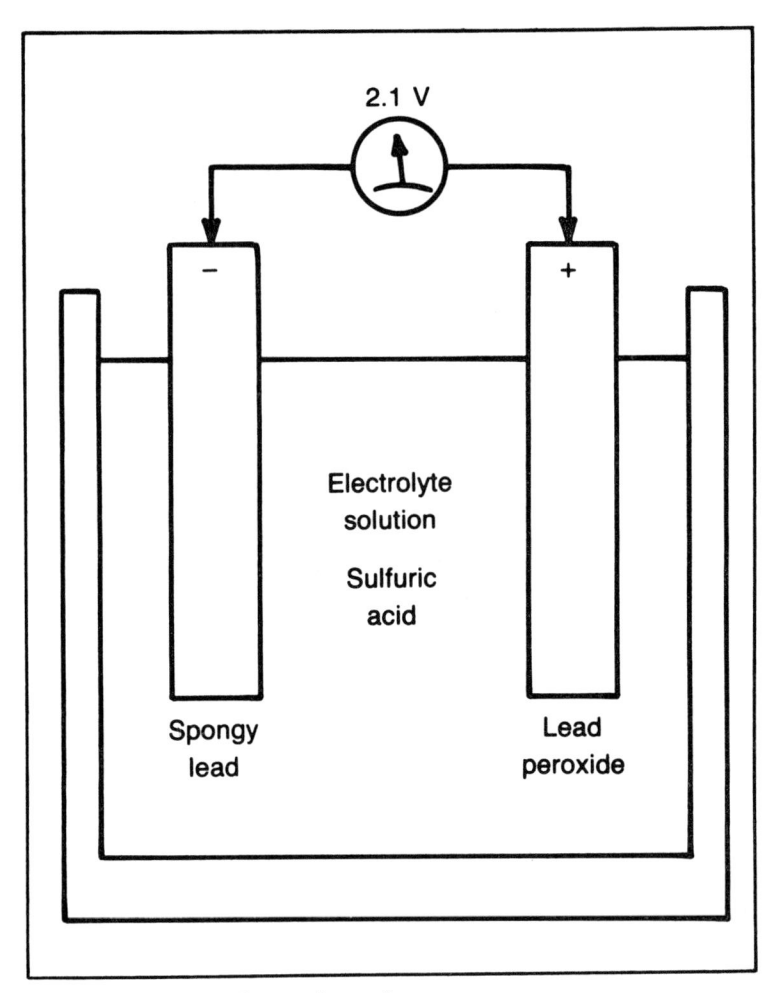

Fig. 5-8. Fully charged secondary cell.

decompose into molecules of water and lead sulfate. The lead-sulfate molecule is deposited on the surface of the lead-peroxide electrode. Figure 5-9 illustrates the chemical transformation taking place when the cell is in use. During this discharge cycle, each electrode is losing its original properties and gaining deposits of lead sulfate. In addition, the concentration of the sulfuric acid is getting weaker.

A completely discharged cell is shown in Fig. 5-10. This illustration shows that a voltage test would indicate approximately 1.5 V across the battery terminals; however, under load the cell could not provide enough current to operate any type of external circuit because of its very high internal resistance caused by the accumulation of lead sulfate on the electrodes.

As we said, the advantage of the secondary cell over the primary cell is that it can be restored or recharged to its original condition. This is accomplished by reversing the current flow and, hence, the chemical action. Figure 5-11 illustrates this charging cycle.

Placing a dc generator across the terminals of the battery causes a reverse chemical action to occur between the electrodes and the electrolyte. The generator's electrical energy provides the force to cause the sulfate deposits to be carried back into the electrolyte solution, increasing the concentration of the acid and returning the

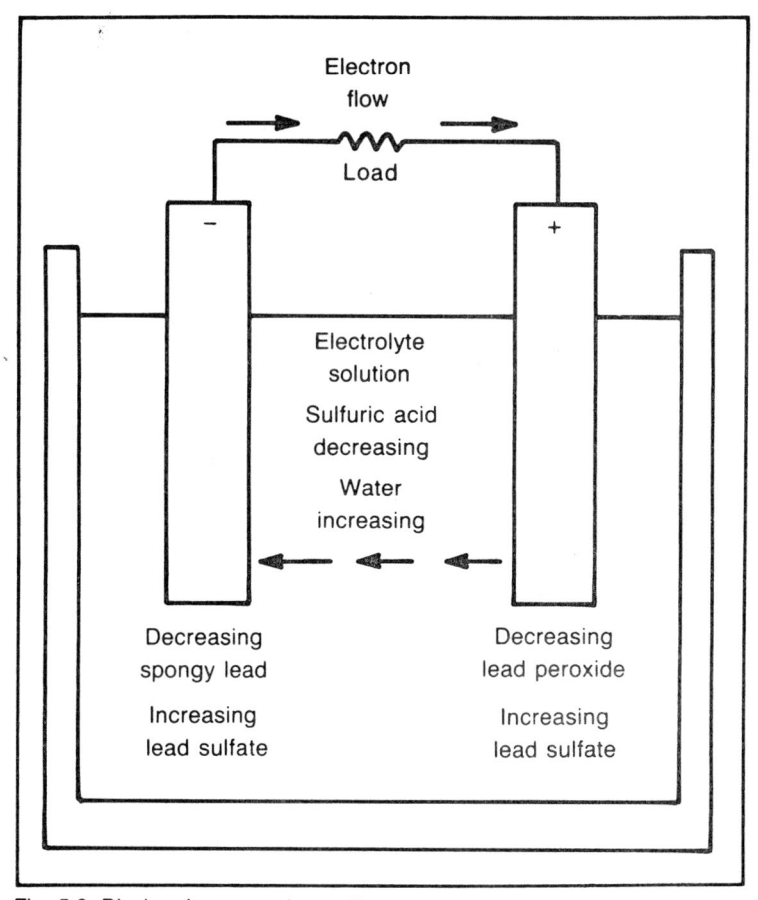

Fig. 5-9. Discharging secondary cell.

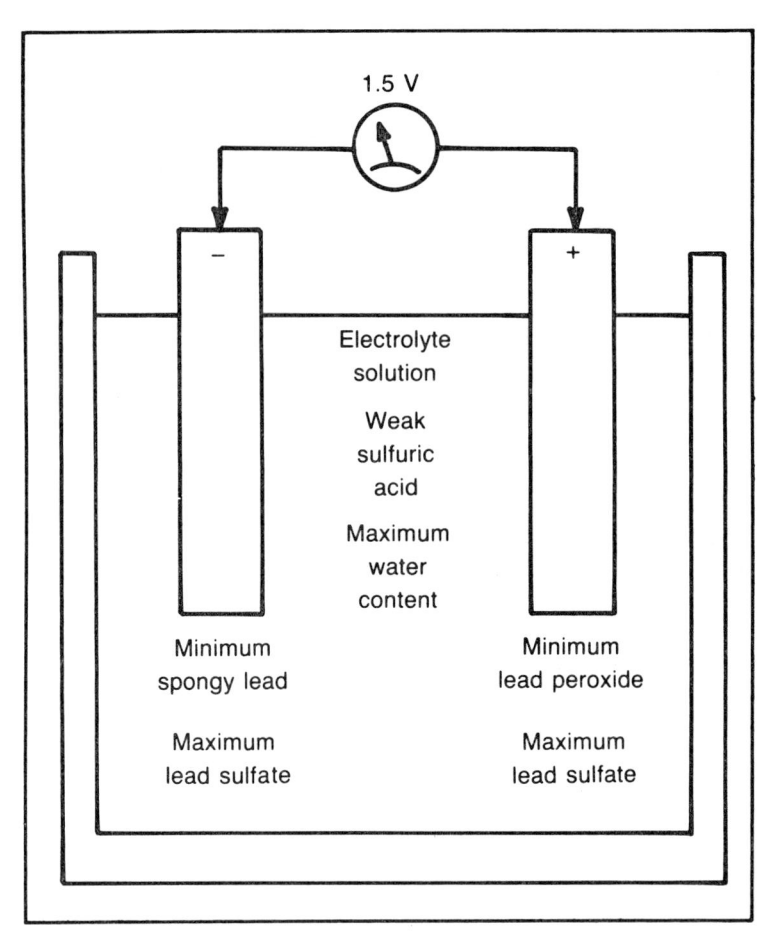

Fig. 5-10. Discharged secondary cell.

electrodes to their original composition. This is accomplished by the electrochemical process known as *electrolysis*.

Secondary Cell Construction

The materials used for electrodes in the lead-acid secondary cell are physically very soft, making it impossible to form them into solid plates. An example of the method used to make these electrodes mechanically rigid is shown in Fig. 5-12. These electrode materials are pressed into a strong lead-antimony grid structure. The chemical action of the electrolyte effects mainly the spongy lead or lead peroxide but not the lead-antimony grid construction.

The electrodes of each cell are arranged in a specific manner so that a large number of electrodes can be placed in a small space within each cell area. Electrode grids are placed side by side and alternate from spongy lead to lead peroxide. The connecting lugs of the spongy-lead grids are all connected in parallel by a negative plate strap, and the lead-peroxide grids are connected in parallel by a positive plate strap. The purpose of this arrangement is to give the cell a large, electrode surface area to increase the current capability of each cell.

An acid-resistant case, such as hard rubber, glass, and some forms of plastic, is used. The case is constructed to provide supporting ribs for the electrodes and space for any sediment (i.e. flakes

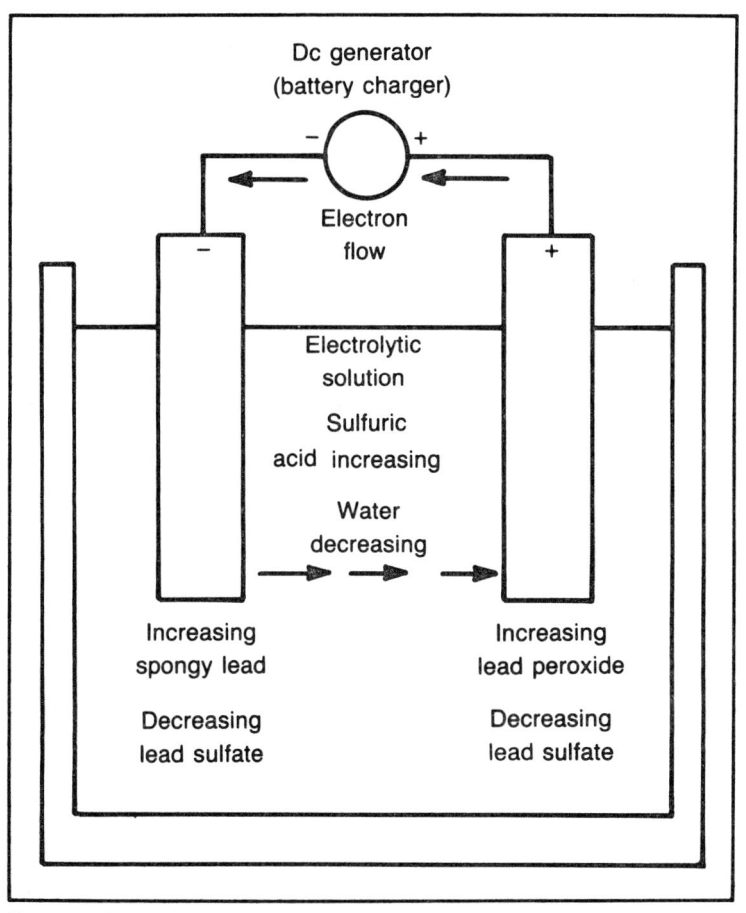

Fig. 5-11. Recharging of a secondary cell.

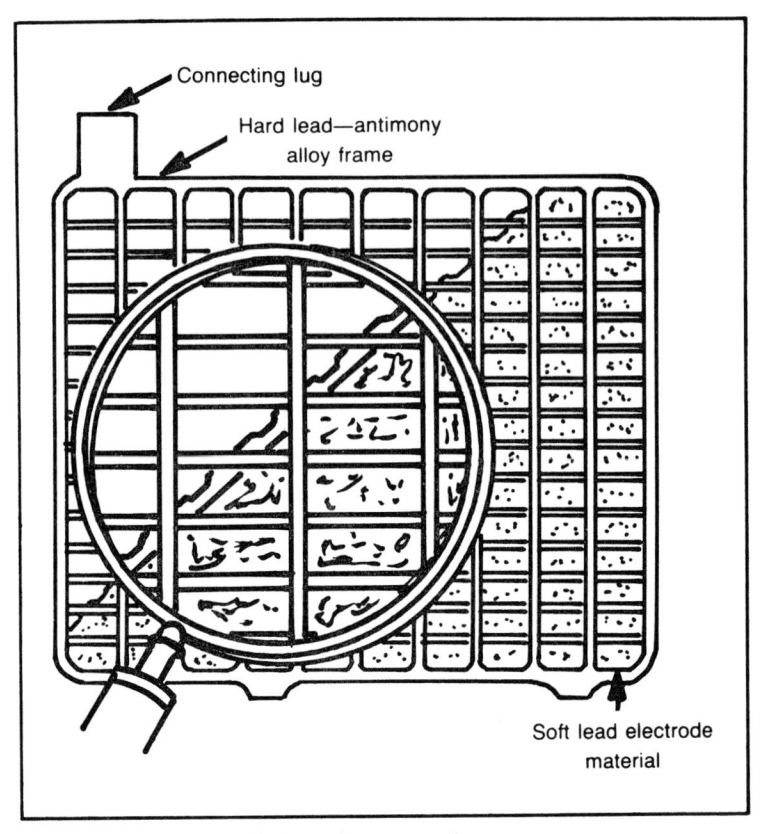

Fig. 5-12. Secondary cell electrode construction.

of the soft lead materials) to collect without danger of internal shorts between electrodes. This is usually what happens to a car battery after operating for three or four years. This accumulation of sediment builds up to the point to where it does short out one of the cells.

There is also adequate spacing provided between electrodes to permit gas formed in the recharging process to escape through the electrolyte filler-cap vent hole. *Caution.* When charging lead-acid secondary cells, hydrogen gas is generated. Hydrogen gas is extremely explosive and always present around the filler-cap vent holes. Keep all open flames away, always provide for adequate ventilation when charging cells, and be extremely careful when getting a jump start when your battery is discharged.

Secondary cells commonly used have six cells and provide an output voltage of 12 V. Figure 5-13 shows the external method of connecting the individual cells to form a 12-V battery.

Ratings of Secondary Cells

There are three factors related to the rating of secondary batteries

- The voltage output
- The current capability
- Battery capacity

The voltage output of any battery depends on the number of cells in series, and the current capability depends on the electrode surface area and the electrolyte quantity.

Battery capacity depends on the amount of current delivered and the time span for which this current can be delivered. This rating of battery capacity is commonly expressed in terms of ampere-hours (Ah). For example, if a battery is capable of delivering 10 A of current for a period of three days, the battery's capacity would be 720 Ah. When a battery manufacturer gives the capacity in ampere-hours, the time and amount of current the battery will deliver is equal to any of the submultiples of the ampere-hour rating. Thus, a 720-Ah battery can deliver 7.2 A for 100 h, etc.

For rapid discharge rates the full battery capacity may not be immediately available because bubbles of oxygen gas on the lead-peroxide plates temporarily reduce the effective grid area.

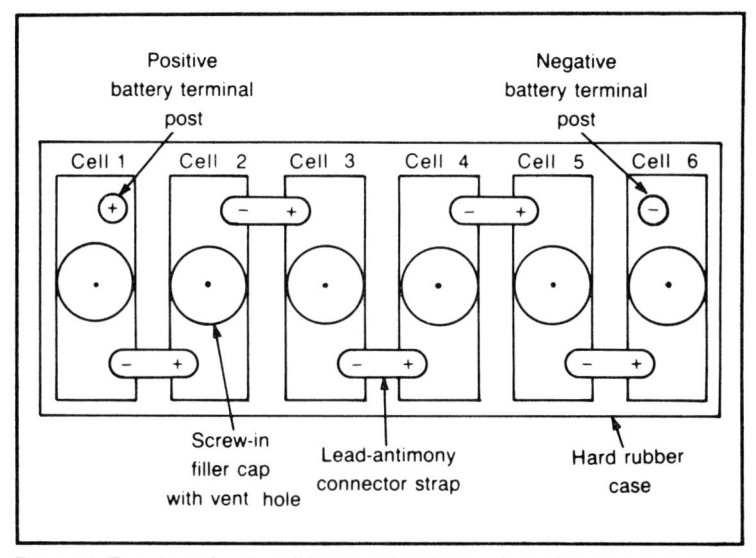

Fig. 5-13. Top view of a six-cell battery showing method of series connection.

Methods of Charging

Secondary batteries can be recharged by two main methods.

- Constant-current method
- Constant-voltage method

The electrical energy required to recharge a secondary battery must come from a dc source. In motor vehicles a dc generator or alternator mechanically attached to the engine supplies this direct current to recharge the battery. In other applications, similar devices are used to provide the necessary voltage and current.

Constant-Current Method

The constant-current method is used to charge single batteries or several batteries connected in series. This method requires a dc source with a voltage output greater than the voltage output of a single battery of a group of batteries. An example of constant-current charging is given in Fig. 5-14. Notice that the positive dc generator terminal is connected to the positive battery terminal through the variable resistor and ammeter, and the negative dc generator terminal is connected to the negative battery terminal. This connection gives the desired current reversal through the electrolyte and the electrodes of the battery. When the battery is being charged, the dc generator is adjusted to produce a voltage slightly higher than the internal voltage of the battery, and the charging current is set at a value between 5 and 10 A. Both of these adjustments are controlled by the variable resistor in the circuit.

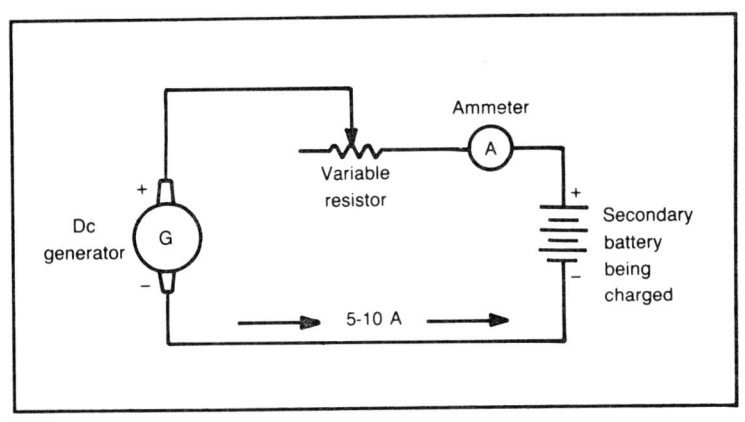

Fig. 5-14. Constant-current charging of a secondary battery.

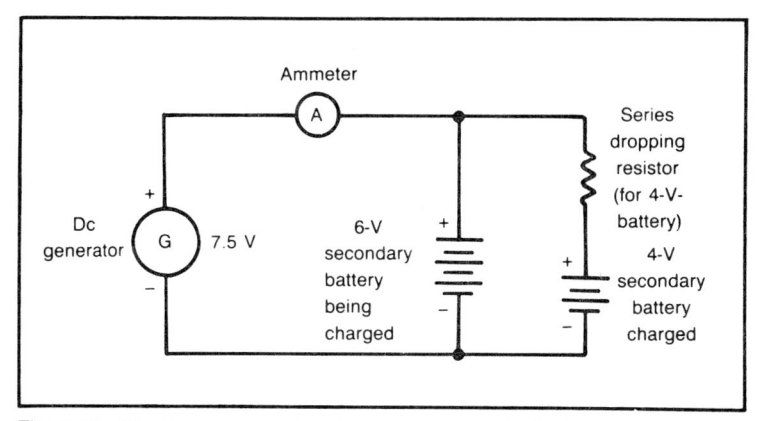

Fig. 5-15. Constant-voltage charging of secondary batteries.

This method is a relatively slow process and requires from 20 to 30 h to fully charge a battery.

This slow charge method is commonly termed a *trickle charge*. It has the advantage of keeping internal battery temperatures at a low level, thereby reducing the danger of heat (caused by passing a current through the high resistance of a discharged cell) that would make the plates buckle and internally short the battery.

Constant-Voltage Charging

Constant-voltage charging is illustrated in Fig. 5-15. This method uses a dc source with a constant voltage output. An ammeter is used to monitor the circuit current. The polarity of the dc source is matched to the polarity of the battery terminals. If charging more than one battery, place the additional batteries in parallel. Figure 5-15 shows that it is possible to charge two batteries of different voltages at the same time as long as a series dropping resistor is placed in the circuit of the battery with the lower potential.

When a discharged battery is placed in a constant-voltage charging circuit, the dc source will automatically furnish a charging current of approximately 5 to 30 A, depending on whether it is impressed across one battery or divided among two or more batteries. As the battery begins to take on a charge, the terminal voltage will increase, causing the current delivered by the generator to decrease. As the battery becomes more fully charged, less current is required to charge it further, and the current in a constant-voltage charging setup tapers off during the charging period. This type of circuit is sometimes called a *taper charger* because of the gradual decrease in applied current.

The constant-voltage method is fast compared to the constant-current method and requires approximately 6 to 8 h to fully charge a battery.

The rates at which batteries should be charged are usually specified by the manufacturer. These rates should be faithfully observed to prevent damage to the cells. If a battery is charged at too high a rate, the electrolyte will bubble violently and excessive amounts of hydrogen and oxygen gases will be formed. An excessive amount of gas in a confined area could cause an explosion. It is common practice to loosen or remove the electrolyte filler caps during the charging process because the vent-hole spacing is too small to permit generated gases to escape.

In addition, when batteries are being charged, at normal or fast rates, the electrolyte level must be properly maintained by adding distilled water to assure complete immersion of the grids.

Electrolyte and Testing

The electrolyte in a lead-acid secondary cell undergoes a chemical change while the battery is being discharged (changing back to distilled water) and while it is being charged (changing back to sulfuric acid). To understand the electrolyte and what it has to do with testing, we will now cover its composition.

Electrolyte. The electrolyte solution of a fully charged battery is composed of approximately 27 parts of acid and 73 parts of water. This ratio of acid and water can be determined by measuring the weight of a given volume of the solution. By comparing this weight with the weight of an equal volume of the solution of a fully charged battery, we can determine the condition of the battery charge.

The method used to determine the percentage of acid in the electrolyte is termed the *specific gravity reading*. Specific gravity is the ratio of the weight of a given volume of any liquid to the weight of an equal volume of water. Water is the most plentiful liquid on the earth and is given the value of 1.000. You can find the specific gravity of pure sulfuric acid by using the following formula.

$$SG = \frac{\text{WT. of 1 ft}^3 \text{ sulfuric acid}}{\text{WT. of 1 ft}^3 \text{ of water}}$$

$$= \frac{114.3 \text{ lb solidus ft}^3}{62.43 \text{ lb ft}^3}$$

$$\text{SG} = 1.830$$

Because the specific gravity of water is 1.000 and the specific gravity of pure sulfuric acid is 1.830, the specific gravity of the electrolyte in a lead-acid cell (27 parts acid and 73 parts water falls somewhere between the two. The specific gravity of a fully charged lead-acid cell is between 1.270 and 1.380.

Testing. To rapidly measure the specific gravity of a cell's electrolyte, we use an instrument called a *hydrometer*. The hydrometer operates on the principle of physics discovered by Archimedes, which is that a floating body will sink deeper into a light liquid than it will into a heavy liquid. Figure 5-16 shows a hydrometer. It contains a weighted glass bulb with a scale calibrated by its manufacturer. This bulb floats freely inside a glass cylinder when a liquid is drawn into the cylinder. The electrolyte is drawn into the cylinder by compressing and then releasing the rubber squeeze bulb with the rubber tip under the liquid's surface. A perforated rubber stopper at the top of the glass cylinder prevents the glass float form being drawn into the bulb. When this electrolyte solution is drawn into the cylinder, the weighted bulb floats freely, and the specific gravity of the electrolyte can be read directly on the float scale. It is important for the hydrometer to be held in a perfectly vertical position with the float bulb floating freely in order to get a true reading. The specific gravity of an electrolyte varies with temperature. Changes in temperature affect the volume of every liquid; if the temperature of a liquid goes up, its volume increases; when the temperature goes down, its volume decreases. To compensate for this, the manufacturer provides a chart with the necessary amount to add or subtract form the hydrometer reading.

The hydrometer is a fairly accurate instrument for testing the condition of a battery. The most accurate method, however, is a combination of a hydrometer check and open- and closed-circuit voltage checks. This combination will indicate the true condition of a battery.

The battery manufacturer furnishes instructions that should be followed in battery maintenance and the maximum and minimum specific gravities for both hot and cold climates.

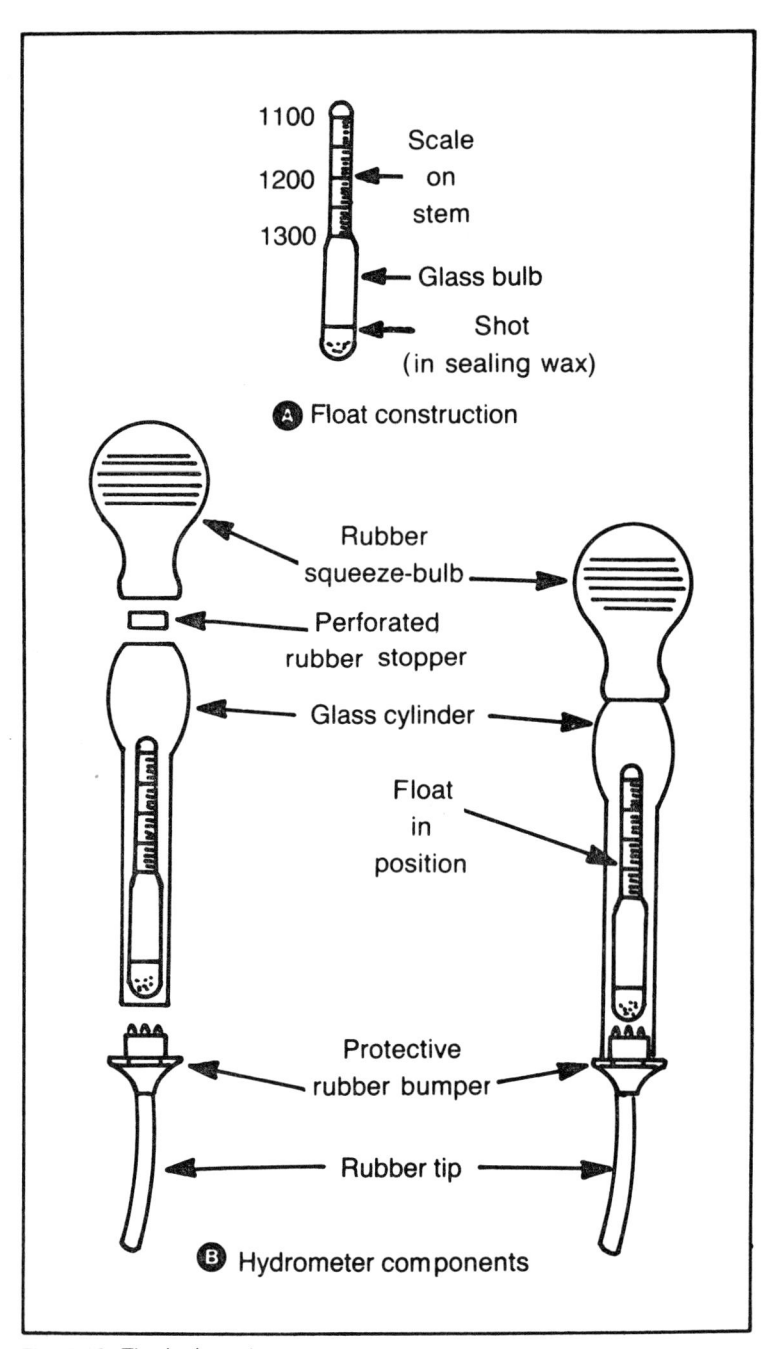

Fig. 5-16. The hydrometer.

Care of Secondary Cells

A battery that has been properly maintained will be much more reliable and have a longer useful life than one that has not. You should remember the following points:

■ *Never* let a battery remain discharged for any appreciable length of time. By recharging the battery immediately, you prevent sulfate deposits from forming hard white crystals that close the pores of the very porous lead electrode materials and will eventually destroy them.

■ Keep the battery connections and terminals free of the copper salt (green) deposits by washing them with a mixture of baking soda and water or ammonia and water. These deposits are a form of corrosion caused by spilling or seeping electrolyte solution and are very detrimental to battery life. If allowed to form, they may create a high-resistance path for current flow and thereby discharge the battery.

■ Always maintain the proper level of electrolyte with distilled water. The proper level is approximately 3/8 in. above the plates. If you do not, the plates will be exposed to air and rapid sulfation and destruction will occur.

■ In cold weather keep batteries as close to full charge as possible, because the freezing point of the electrolyte depends on its specific gravity, and, in turn, the specific gravity depends on the charged condition of the battery. If the specific gravity of the electrolyte drops, the freezing point of the electrolyte will rise (Fig. 5-17).

■ *Never* add extra acid to a cell. If you know electrolyte has been spilled, replace only the amount of acid spilled. Excessive acid or electrolyte is wasted because it will only raise the specific gravity of the electrolyte and give a false reading. It will recharge the cell. Only a dc charging current can chemically change the solution and charge a battery.

QUIZ

1. What is the main difference between a primary cell and a secondary cell?
2. Series-connected cells provide more _____ .
3. Parallel-connected cells provide more _____ .
4. When secondary batteries are recharged _____ , gas is generated.
5. A 500-A battery can supply 50 A for _____ hours.

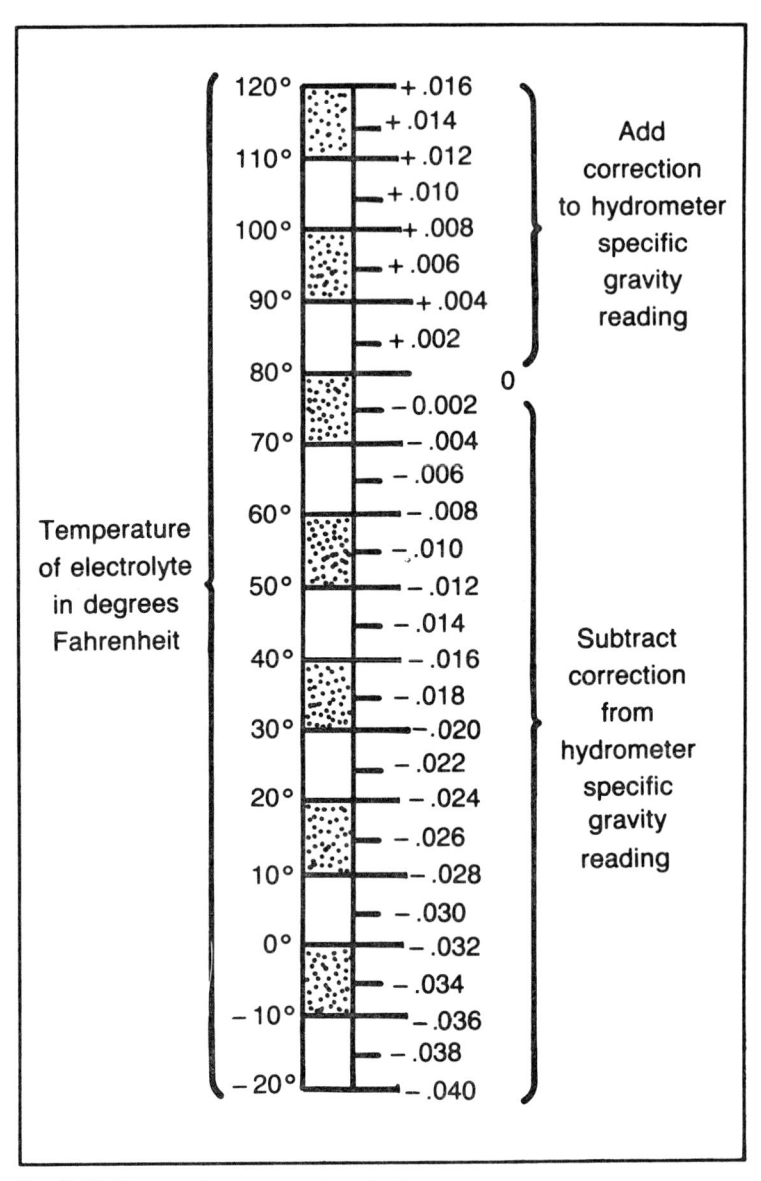

Fig. 5-17. Temperature conversion chart.

6. Two methods for recharging secondary batteries are _____ and _____ .

7. Charging a secondary battery at too fast a rate can cause _____ .

8. The liquid solution in a secondary battery is called the _____ .

9. A measure of the _____ of the electrolyte solution is a method of checking a battery's condition.

10. The instrument used to check specific gravity is called a _____ .

Chapter 6

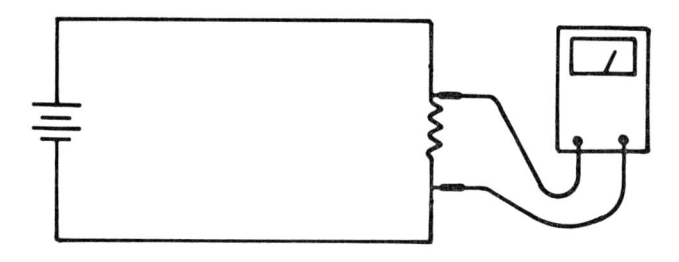

Resistors

ALTHOUGH WE BASICALLY COVERED THE TWO PRINCIPAL uses of resistors in Chapter 3, the resistor is so important that it deserves a special chapter.

A resistor is used to

- Control and limit the amount of current in a circuit
- Develop a voltage (IR) drop

They have two important physical characteristics:

- Their ohmic value in ohms
- Their power rating in watts

Providing proper voltages and currents are the key elements in designing and building any type of electrical and electronic equipment. The use of resistors, therefore, is very widespread: if the components used in all electrical and electronic equipment were counted, resistors would probably outnumber all others.

TYPES OF RESISTORS

Resistors can be grouped into three main categories:

- Carbon-composition resistors
- Wirewound resistors

■ Carbon-film resistors

Each of these can be fixed, tapped, or variable.

Carbon-Composition Resistors

Carbon-composition resistors are the most common. A carbon-composition resistor is simply a carbon compound held together with cement. They are made with numerous ohmic values, although 1/2-W, 1-W, and 2-W sizes predominate. The relative physical size of carbon-composition resistors is shown in Fig. 6-1.

Carbon-composition resistors cannot be made as accurately as wirewound or carbon-film resistors; therefore they are not used in circuits requiring close resistance tolerances.

Wirewound Resistors

A wirewound resistor is made by winding a resistance wire on an insulated support such as a ceramic-type rod. Wirewound resistors are found with wattage ratings from 3 to 4 W up to some very high values.

In some equipment you will find tapped wirewound resistors that look like the resistor shown in Fig. 6-2. Its total resistance is 1000 ϕ. This is the resistance that you would measure with an ohmmeter between terminals A and D. However, there are two taps (B and C) on the resistor. The resistance between A and B is 250 ϕ, that between B and C is 500 ϕ. This type of resistor is usually quite large and heavy. Therefore, it is usually mounted on a chassis by brackets to prevent its breaking loose by jarring or bouncing.

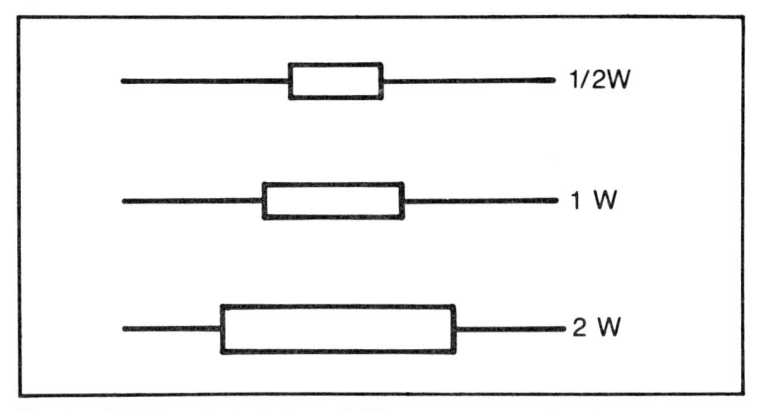

Fig. 6-1. Relative physical sizes of different power-rated resistors.

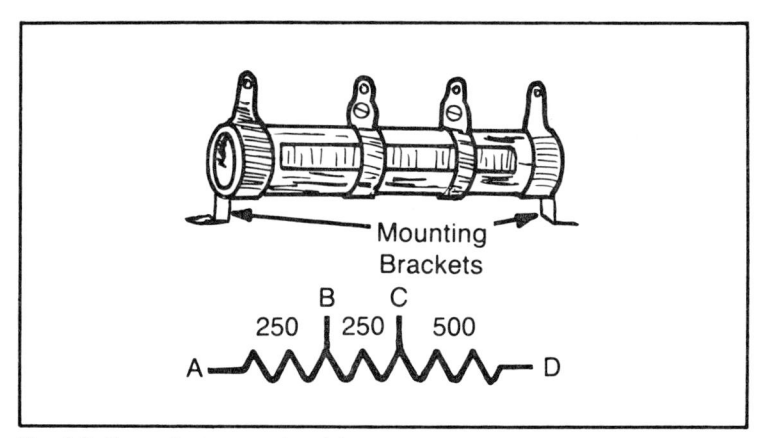

Fig. 6-2. Tapped wirewound resistor.

Because the resistor is made from a coil of wire, it takes on some characteristics of an inductor. Chapter 9 explains how these windings can be made to minimize this effect. This type of winding is called noninductive.

Variable Resistors

Figure 6-3 illustrates different types of variable resistors. The resistor in Fig. 6-3A is a wirewound type called a *rheostat*. Notice that it has only two terminals. Another rheostat is shown in Fig. 6-3B. It can be a wirewound control similar to that in Fig. 6-3A, or it may have a carbon element with a slider that rotates along the element to provide the required resistance between the two terminals. The variable resistor shown in Fig. 6-3C is called a *potentiometer*. It has three terminals. The resistance between the two

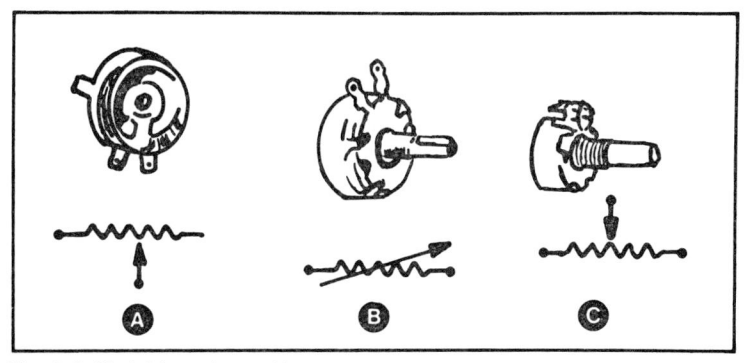

Fig. 6-3. Variable resistors (rheostats, potentiometers).

outside terminals remains constant, and the slider moves on the resistance element so that the resistance between the center terminal and the other two may be varied. A control of this type may be wirewound carbon.

Most potentiometers (called "pots") are high-resistance units with carbon controls. In some applications, such as color TV receivers, very low-resistance potentiometers with wirewound controls are used. Potentiometers are found in all types of electrical and electronic equipment.

RESISTOR VALUES

The resistor is the most common component used in electronic equipment. A small radio has about 15 to 20 resistors, black and white TV receiver about 50 to 75, and a color TV almost twice as many. Circuit cards can have hundreds and thousands of resistors. In general, you can assume that any type of electronic equipment will have up to 40 resistors for each integrated circuit used.

Tolerances

Resistors are manufactured with distinct tolerances. The most frequently encountered resistor is the molded carbon-composition resistor. It is manufactured with three main tolerance values: 5 percent, 10 percent, and 20 percent. The tolerance figure indicates how much the resistor may deviate from its indicated resistance value. For example, a 100-ϕ resistor with a 5-percent tolerance can have a resistance between 95 and 105 ϕ.

The closer the tolerance of a resistor, the more expensive it is. Thus a 5-percent resistor is more expensive than a 10-percent resistor, and a 10 percent resistor is more expensive than a 20-percent resistor.

Electronic Industries Association

The Electronic Industries Association (EIA) has developed standard carbon-composition resistor values. You cannot simply buy any size resistor you like, you must buy a standard value. The standard values are arranged so that you can get a resistor within 5 percent of any required value. For example, if you wanted a 53,000-ϕ resistor, you couldn't buy one. You can buy a 56,000-ϕ resistor. Standard EIA carbon-resistor values are shown in Table 6-1. All are available with a 5-percent tolerance; the values in bold type are also available in 10-percent tolerances. You normally can-

Table 6-1. EIA Carbon-Composition Resistor Values.

Ohms	Ohms	Ohms	Ohms	Ohms	Ohms	Ohms	Ohms	Megs	Megs
0.24	1.1	5.1	24	110	510	2400	11 k	51 k	240 k	1.1	5.1
0.27	1.2	5.6	27	120	560	2700	12 k	56 k	270 k	1.2	5.6
0.30	1.3	6.2	30	130	620	3000	13 k	62 k	300k	1.3	6.2
0.33	1.5	6.8	33	150	680	3300	15 k	68 k	330 k	1.5	6.8
0.36	1.6	7.5	36	160	750	3600	16 k	75 k	360 k	1.6	7.5
0.39	1.8	8.2	39	180	820	3900	18 k	82 k	390 k	1.8	8.2
0.43	2.0	9.1	43	200	910	4300	20 k	91 k	430 k	2.0	9.1
0.47	2.2	10	47	220	1000	4700	22 k	100 k	470 k	2.2	10
0.51	2.4	11	51	240	1100	5100	24 k	110 k	510 k	2.4	11
0.56	2.7	12	56	270	1200	5600	27 k	120 k	560 k	2.7	12
0.62	3.0	13	62	300	1300	6200	30 k	130 k	620 k	3.0	13
0.68	3.3	15	68	330	1500	6800	33 k	150 k	680 k	3.3	15
0.75	3.6	16	75	360	1600	7500	36 k	160 k	750 k	3.6	16
08.2	3.9	18	82	390	1800	8200	39 k	180 k	820 k	3.9	18
0.91	4.3	20	91	430	2000	9100	43 k	200 k	910 k	4.3	20
1.0	4.7	22	100	470	2200	10 k	47 k	220 k	1 meg	4.7	22

111

not buy 20-percent resistors for replacement purposes, but either 5-percent or 10-percent resistors are better substitutes.

THE RESISTOR COLOR CODE

A standard color-coding system has been developed for identifying resistor values. This system includes both ohmic values in ohms and a tolerance in percent. Figure 6-4 shows a four-band color-code system. Reading from left to right, the first band indicates the first significant figure, the second band the second significant figure, the third band the number of zeros following.

The tolerance values for the fourth band are

Silver   10 percent
Gold   5 percent
No color 20 percent

Note the examples given in Fig. 6-4.

TEMPERATURE COEFFICIENT

A very noticeable characteristic of carbon-type resistors is that they change resistance with changes in temperature. Carbon, unlike most conductors, has what is called a *negative temperature coefficient*. This means that as carbon is heated, its resistance decreases. Copper, on the other hand, has a *positive temperature coefficient*. If a copper conductor is heated, its resistance will increase.

In some cases due to the resistor construction and the type of material used as a binder a carbon resistor may actually have a small positive temperature coefficient.

Wirewound resistors usually have positive temperature coefficients. Obviously, the ideal situation, in general, is to use resistors with zero temperature coefficients because resistance changes due to temperature often produce undesirable results.

In some special applications we want a resistor whose value will change with temperature. This type of resistor is manufactured with special characteristics. Two types are thermistors and varistors.

Thermistors

A *thermistor* is a resistor whose value varies with changes in temperature. The material used in a thermistor is a form of semiconductor material very similar to that used in transistors. Thermistors

Color	Code Digit 1st	Code Digit 2nd	Multiplier
Black	0	0	1
Brown	1	1	10
Red	2	2	100
Orange	3	3	1,000
Yellow	4	4	10,000
Green	5	5	100,000
Blue	6	6	1,000,000
Violet	7	7	10,000,000
Gray	8	8	100,000,000
White	9	9	1,000,000,000
Gold	—	—	.1
Silver	—	—	.01

Tolerance
Gold ± 5 percent
Silver ± 10 percent
No band ± 20 percent

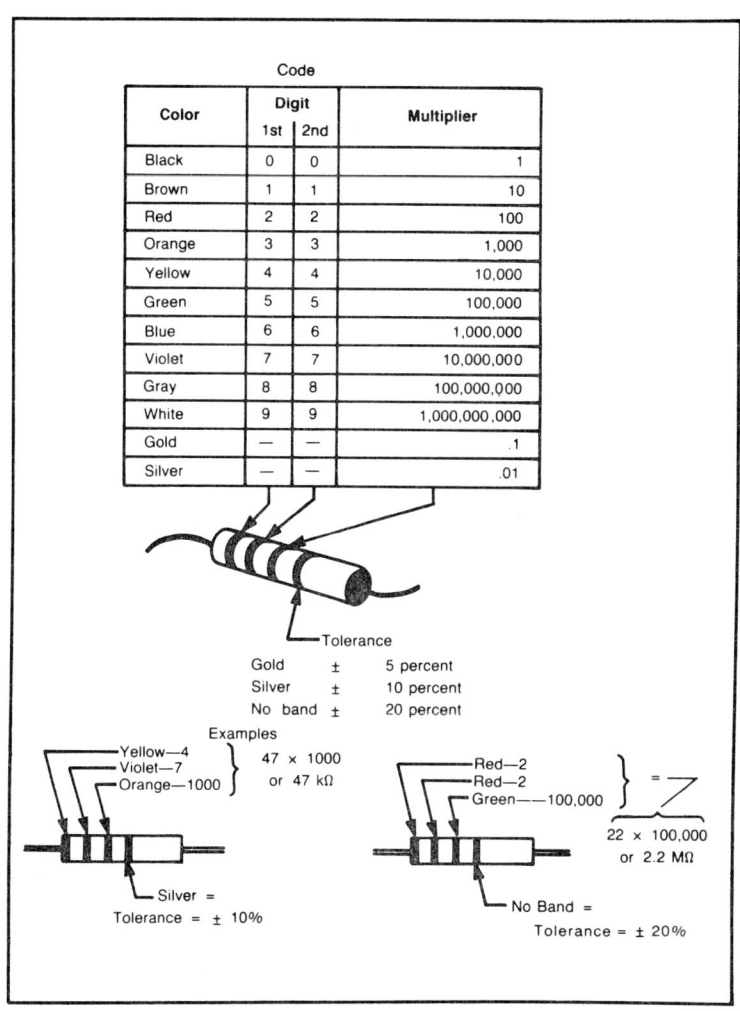

Examples

Yellow—4
Violet—7
Orange—1000
47 × 1000 or 47 kΩ
Silver =
Tolerance = ± 10%

Red—2
Red—2
Green—100,000
22 × 100,000 or 2.2 MΩ
No Band =
Tolerance = ± 20%

Fig. 6-4. Resistor color codes.

have a negative temperature coefficient. The amount of change with a given change in temperature depends on the type of material used in its construction. See Fig. 6-5.

Thermistors are often used in circuits where there may be large surges of current when the equipment is first turned on. For example, in the power supply of some TV receivers after the set is first turned on, a very high surge of current will flow for a very short time. If this is allowed to happen for long, these high-current surges can damage circuit components. A thermistor can limit this

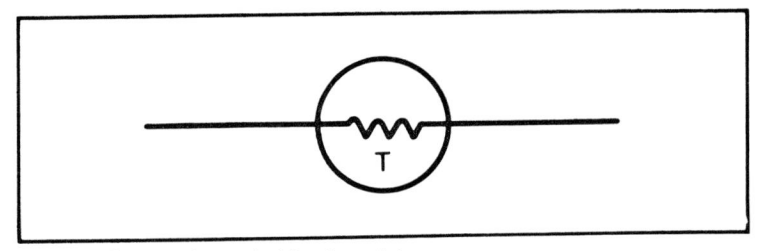

Fig. 6-5. Schematic symbol for thermistor.

current to a safe value. As the equipment starts to operate, the current flowing through the thermistor causes it to heat, so its value drops rapidly. When the equipment reaches its operating temperature and is ready to function, the resistance of the thermistor has dropped to such a low value that it has very little effect on the performance of the equipment, but it has performed its function in limiting initial current surges. A thermistor of this type will have a resistance of perhaps 100 ϕ when cold and only a few ohms when its normal operating temperature is reached.

Varistors

The thermistor is a temperature-controlled device. The *varistor* is a voltage-controlled device. As shown in Fig. 6-6, the schematic symbol is identical except for the letter "V" at the bottom. The resistance of this device depends on the voltage across it. In other words, as the voltage across the varistor increases, the resistance also increases.

Varistors are used in circuits where there is likely to be a sudden increase in voltage. This would cause a sudden current increase that could cause some undesirable results. With a varistor in the circuit, as the voltage increases suddenly, the resistance of the varistor would increase rapidly and limit the current increase.

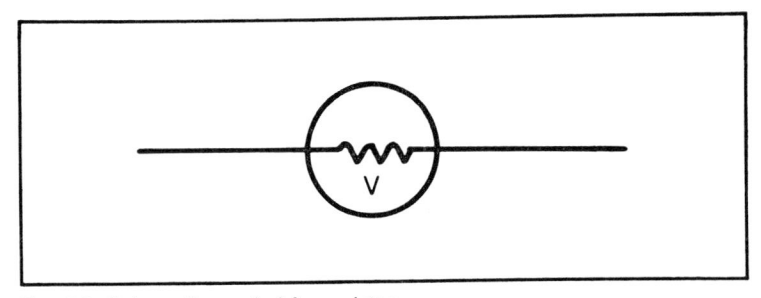

Fig. 6-6. Schematic symbol for varistor.

QUIZ

1. Is a wirewound resistor more accurate than a carbon resistor?
2. Which resistor is physically larger: 1/2 W or 2 W?
3. What are the two main types of variable resistors?
4. A 100-ϕ resistor with a 5-percent tolerance can vary from _____ ϕ to _____ ϕ.
5. What is the value of a resistor marked with green, black, orange, and gold bands?
6. Carbon has a _____ temperature coefficient.
7. A device purposely manufactured with a negative temperature coefficient is a _____.
8. A _____ is a voltage-controlled resistor.
9. Would a 10-W resistor be carbon or wirewound?

Chapter 7

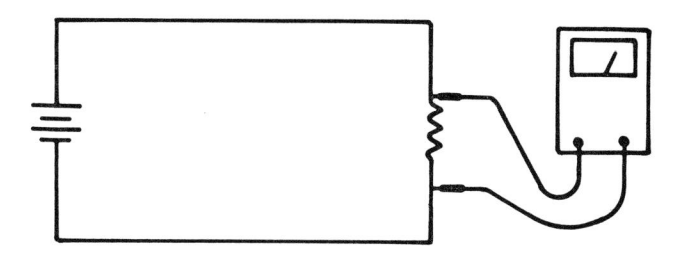

Dc Measuring Devices

EARLY ELECTRICAL EXPERIMENTERS WERE AT A DISADVANtage in explaining their observations because they could not satisfactorily measure actual currents, voltages, and resistances. Over time various instruments called *meters* were developed. They were of three principal types: those that measured current were called *ammeters*; those that measured voltage were called *voltmeters*; and those that measured resistance were called *ohmmeters*. A modern *multimeter,* one of the handiest test instruments ever made, contains all of these meters in one case. This chapter covers the design, construction, and use of each type of meter.

THE AMMETER

An ammeter indicates the quantity of current in an electrical circuit. You must place an ammeter in *series* with the circuit or circuit element through which current is to be measured. Figure 7-1 shows the proper connection of an ammeter.

The ammeter does not actually count the number of electrons passing a particular point in a certain time. Instead, it samples the intensity of the current. But the heart of the ammeter is the meter movement. The most common is the *moving coil*, or *D'Arsonval*, meter movement. A pointer attached to the moving-coil meter movement indicates on a calibrated scale the amount of current passing through the coil (see Fig. 7-2). Meter movements are usually classified according to *meter sensitivity*—that is, the maximum

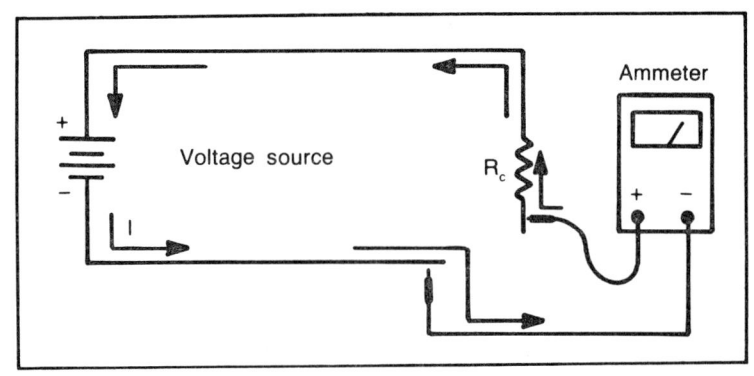

Fig. 7-1. Proper connection of an ammeter.

amount of current required to position the pointer to the full-scale position.

One characteristic of the meter movement is that current must pass through it in the correct direction for proper operation. Figure 7-3 shows a dc ammeter connected to an electrical circuit. R_a and R_c represent the ammeter resistance and the circuit resistance, respectively. When the ammeter is manufactured, R_a is kept as small as possible in order to minimize loading effect. Figure 7-4 illustrates loading effect.

In Fig. 7-4A it is evident from Ohm's Law that with a voltage of 10 V and a circuit resistance of 50 Ω the circuit current is 200 mA. When an ammeter with an internal resistance of 50 Ω is connected to this circuit, the resistance of the ammeter is added to the circuit resistance and current decreases (in this case to 100 mA). See Fig. 7-4B.

Meter movements come with relatively low, fixed sensitivities. For example, in Fig. 7-3 the full-scale deflection is 1 A. This means that, as a test instrument, the ammeter is limited to measuring circuit currents not exceeding 1 A. If we want to measure higher values of current, we must use suitable resistances called meter *shunts*.

Figure 7-5 shows the same electrical circuit and meter movement with the addition of a meter shunt, R_s. Assume the basic meter movement has a full-scale deflection of 1 A. Using Ohm's Law, we can add the proper value of shunt resistor so that the ammeter circuit now allows 5 A to pass before the pointer indicates full-scale deflection. A current 4 A now flows through the shunt, and 1 A flows through the movement. The maximum current-reading ability of the ammeter has been increased by adding the shunt resistance.

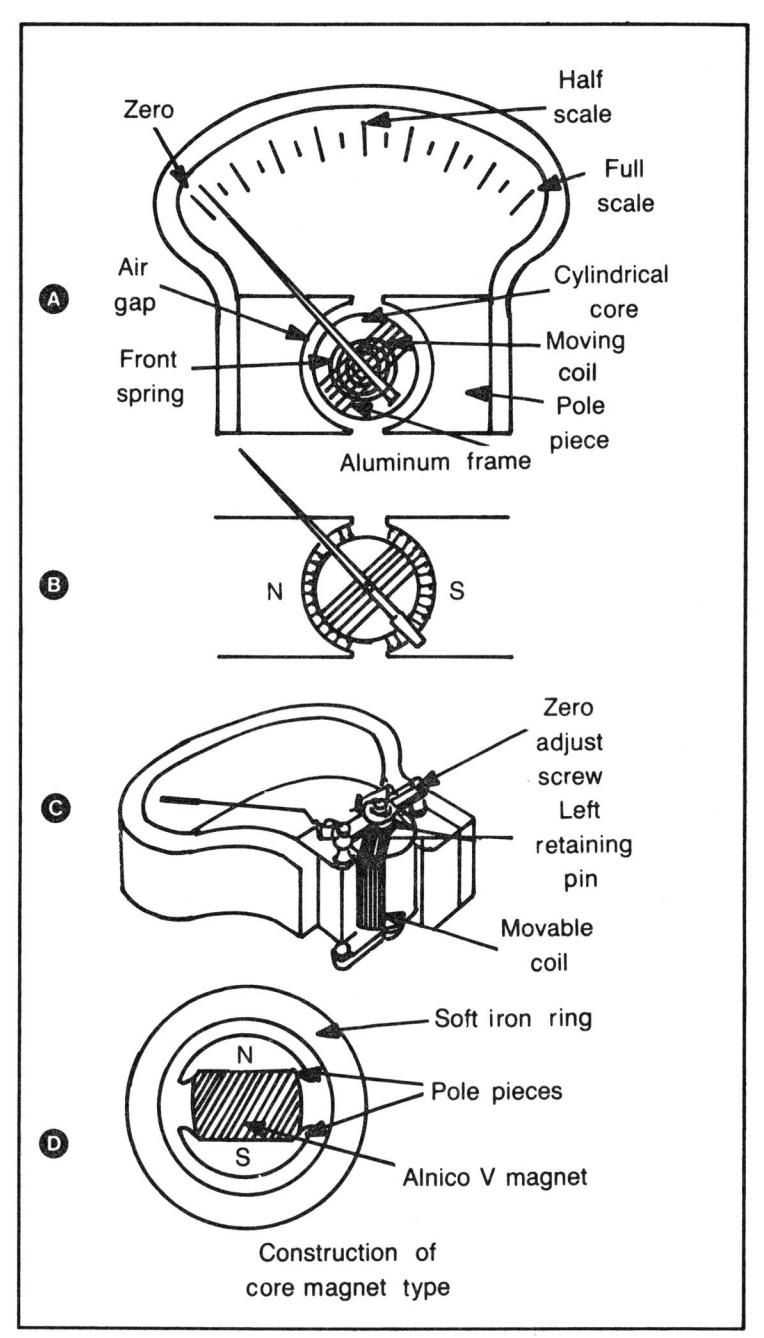

Fig. 7-2. Moving coil or D'Arsonval meter movement.

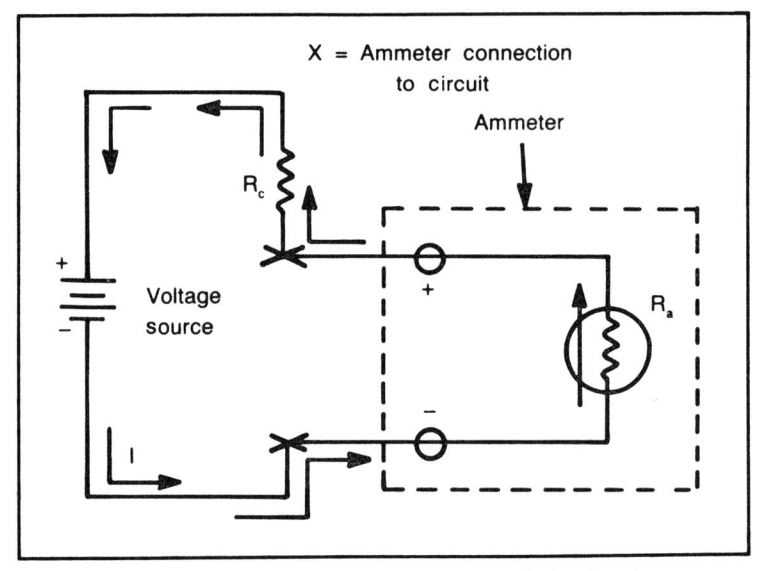

Fig. 7-3. Series effect of ammeter resistance on total circuit resistance.

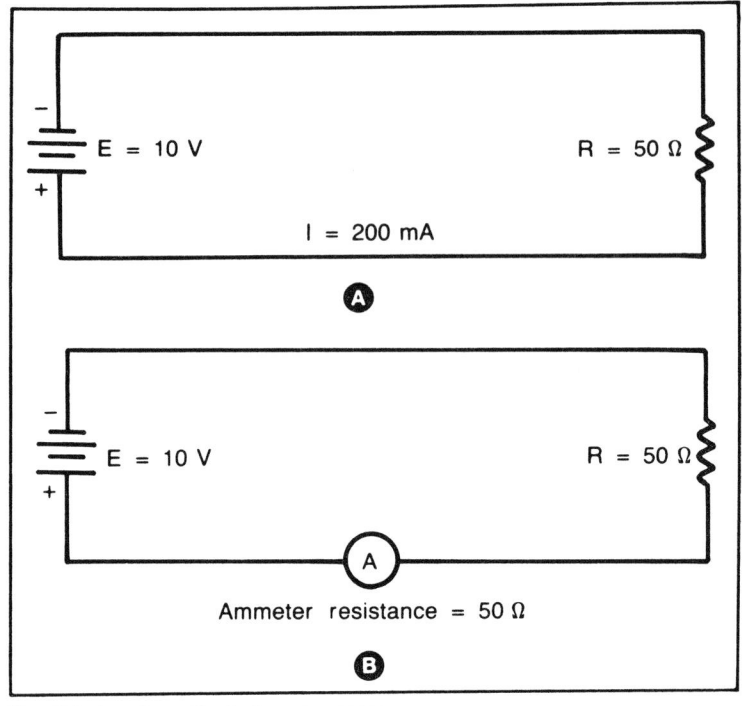

Fig. 7-4. Loading effect of an ammeter.

120

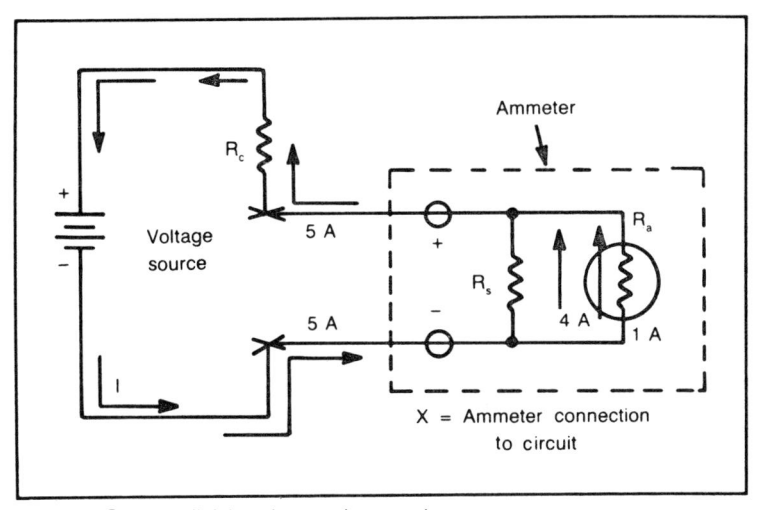

Fig. 7-5. Current division due to shunt resistor.

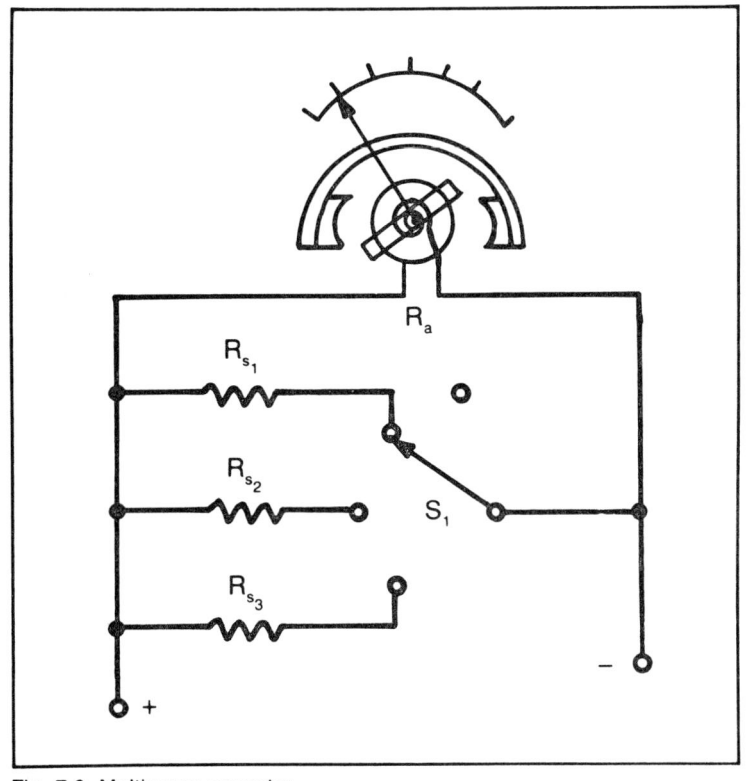

Fig. 7-6. Multirange ammeter.

121

A practical multirange ammeter incorporating different shunt resistors and a switching arrangement is shown in Fig. 7-6.

Ammeter Scales

A typical ammeter scale is shown in Fig. 7-7. The meter face shown represents a multirange of meter with eight possible ranges. Observe the different readings indicated by the needle as it rests on a major scale division. Figure 7-8 shows the needle resting on an intermediate scale division. Intermediate divisions are represented by one-half the value between numbered divisions. By making a very careful comparison between the needle, scale, and table, we can read each of the values listed for the different ranges.

Many times the needle will indicate a point between markings. You will then need to interpolate to get an accurate reading. This is shown in Fig. 7-9.

When reading a meter scale, you should always look directly at the pointer to get an accurate reading. Glancing at the meter face from either side of it produces *parallax* error. Some meters even have a mirror on the scale to correct for this.

Safety Precautions

An actual ammeter is usually multirange with many combinations of shunts and a very low-sensitivity meter movement. It is a delicate and expensive instrument that can be damaged very easily if improperly handled. Dropping or jarring the ammeter may upset the delicate balance of the meter movement. If you choose the wrong scale to measure current flow, excessive current may burn out or destroy the movement. So *always* observe the following precautions:

■ *Never* connect an ammeter across a source of emf.

■ Observe polarity when connecting an ammeter in a dc circuit. In most cases the black meter lead is considered to be the negative lead (−), and the red meter lead is considered to be the positive lead (+).

■ Always connect the ammeter in series with the circuit or circuit element through which current is to be measured. This means opening up the existing circuit.

■ Always use a range large enough to keep the pointer deflection less than full scale. When measuring an unknown value of current always start with the highest range. Then step down until the

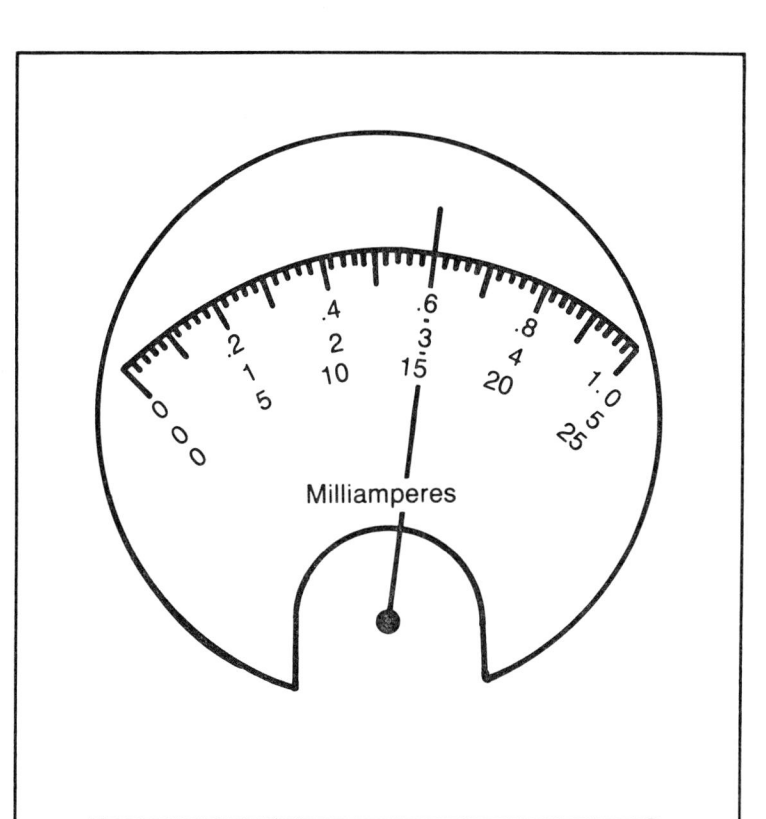

Scale (mA)	Multiplication Factor	Reading (mA)
0-25	none	15
0-250	X 10	150
0-5	none	3
0-50	X 10	30
0-500	X 100	300
0-1.0	none	0.6
0-10	X 10	6.0
0-100	X 100	60.0

Fig. 7-7. Reading ammeter scales with pointer resting on major scale divisions.

Scale (mA)	Intermediate Mark	Reading (mA)
0-25	2.5	7.5
0-250	25.0	75.0
0-5	0.5	1.5
0-50	5.0	15.0
0-500	50.0	150.0
0-1.0	0.1	0.3
0-10	1.0	3.0
0-100	10.0	30.0

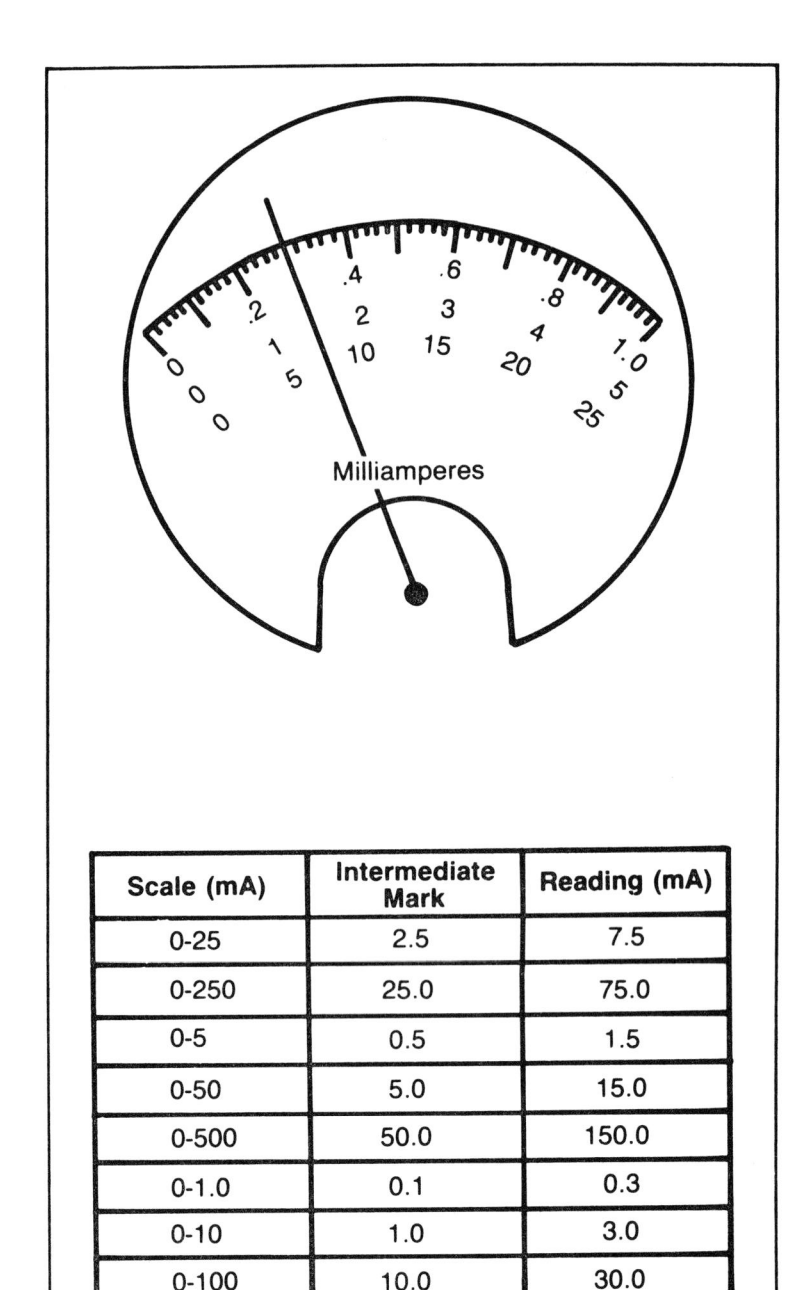

Fig. 7-8. Reading ammeter scales with pointer resting on unnumbered scale position.

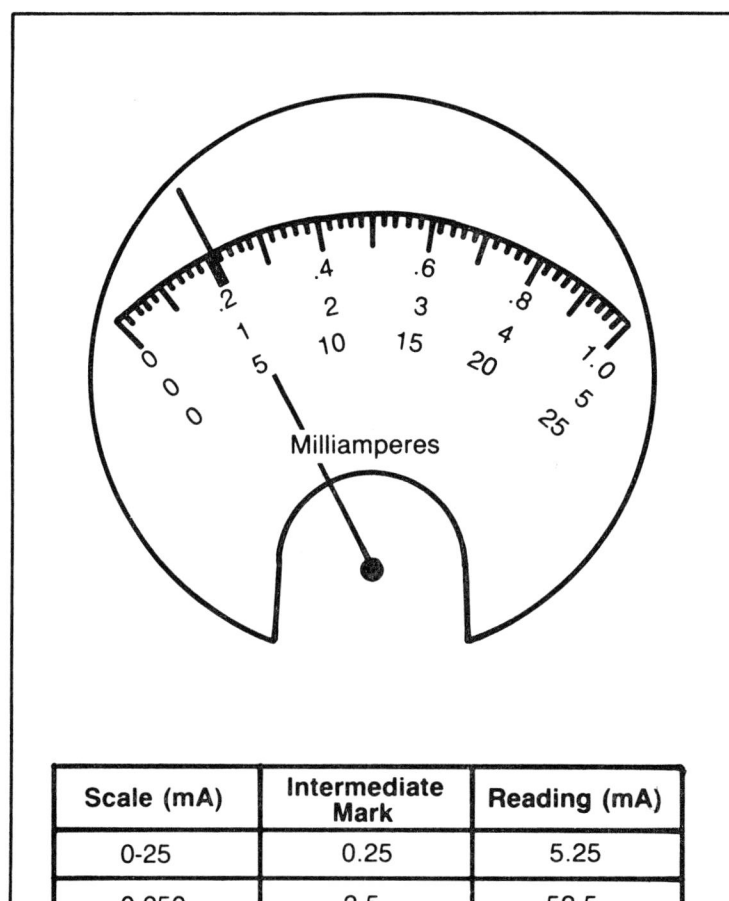

Scale (mA)	Intermediate Mark	Reading (mA)
0-25	0.25	5.25
0-250	2.5	52.5
0-5	0.05	1.05
0-50	0.5	10.5
0-500	5.0	105.0
0-1.0	0.01	0.21
0-10	0.1	2.1
0-100	1.0	21.0

Fig. 7-9. Reading ammeter scales with pointer resting on a fractional part of a scale division.

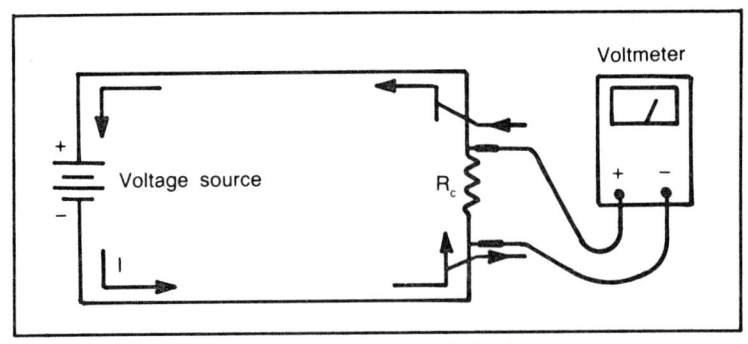

Fig. 7-10. Voltmeter connected to an external circuit.

proper range is reached. (The most accurate readings are found in the mid-scale region of the meter.)

THE VOLTMETER

A voltmeter measures the quantity of voltage in an electrical circuit. To correctly use a voltmeter, place it in *parallel* with the circuit (see Fig. 7-10).

Figure 7-11 shows the internal construction of a voltmeter prop-

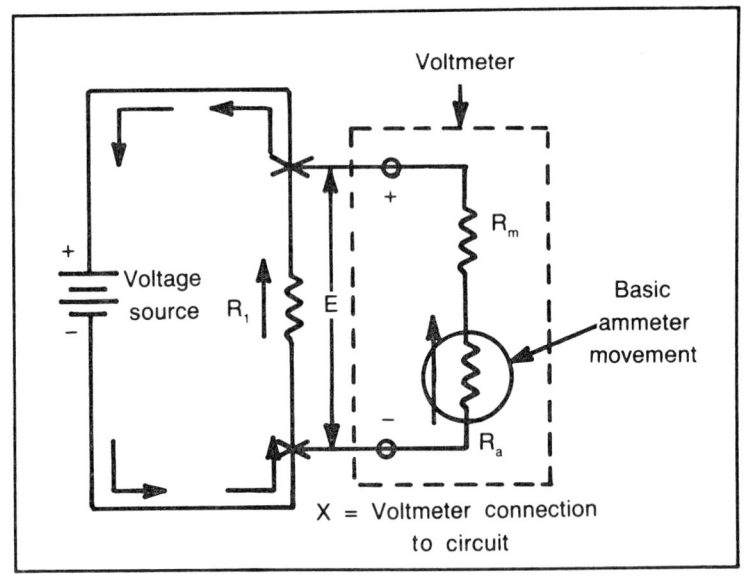

Fig. 7-11. Voltmeter construction showing the use of multiplier R_m connected to an external circuit.

126

erly (enclosed within the dashed lines) connected to an electrical circuit. Note that the voltmeter circuit consists of an ammeter meter movement in series with a resistor, R_m, which is a high-ohmic-value resistor (called a *multiplier* resistor) to reduce the amount of current through the meter movement.

The flow of current through the meter movement causes a deflection of the pointer. By calibrating the scale of the meter movement in volts for a specified current through the moving coil and its series resistor, we can create a complete voltage scale. This scale will extend from zero to the point at which the applied circuit voltage causes enough current to be passed through the meter movement and R_m to produce full-scale deflection. By using several multiplier resistances, we can make a multirange voltmeter, such as the one shown in Fig. 7-12.

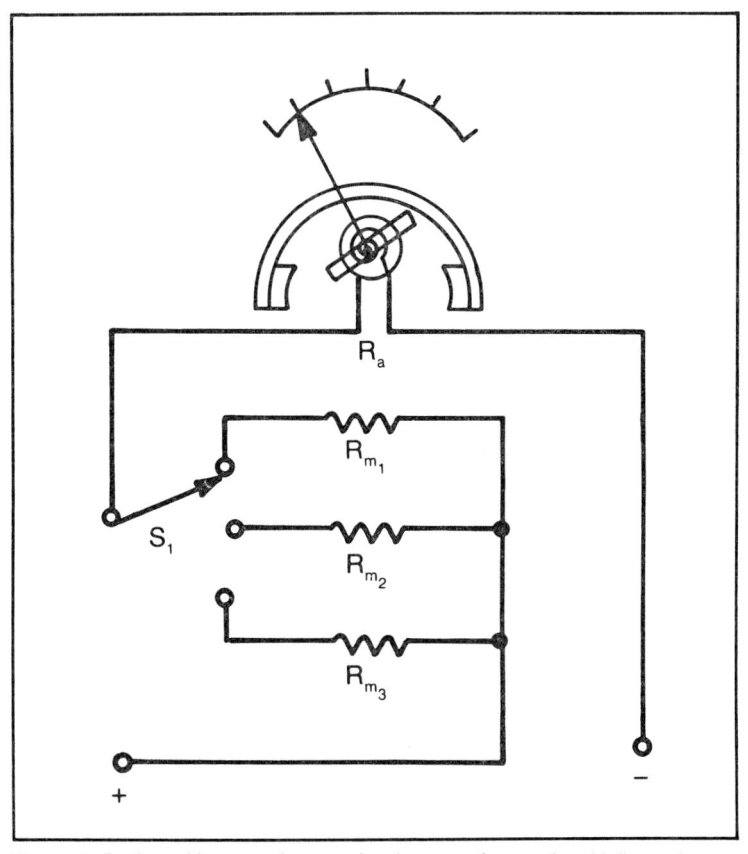

Fig. 7-12. Basic multirange voltmeter showing use of several multiplier resistors.

Voltmeter Sensitivity

The ability of a voltmeter to measure circuit voltages accurately is determined by its sensitivity. The sensitivity of a voltmeter (in ohms per volt) is expressed differently from that of the ammeter because it is determined by the ratio of total internal meter resistance to the maximum amount of voltage on its selected scale. The common sensitivities of different voltmeters range from 1,000 to 20,000 ohms per volt (Ω/V). The higher the sensitivity, the greater the accuracy of the meter. This is true because the current required to operate the meter movement is much smaller for high-sensitivity voltmeters.

Loading Effect of Voltmeters

Because a voltmeter must be connected in parallel with a circuit component, the operation of the circuit may be effected, and this could produce inaccurate readings. The inaccuracy is due to the loading effect of the meter on the circuit.

Figure 7-13 illustrates this effect. We would expect that when the voltmeter is placed across resistor R_2 it would read 200 V, but, depending on the sensitivity of the voltmeter, it could give readings other than 200 V. Notice that the meter used has a rating of 1000 Ω/V. On the 0-250 scale the meter circuit resistance will be 250,000 Ω. With the voltmeter connected across R_2 a parallel circuit is formed, making an effective resistance of 111,000 Ω. This is shown in Fig. 7-14. Because the effective circuit resistance is decreased, it causes the circuit current to increase to 1.42 mA, and the voltage drop is now 158 V. This is the voltage that will be indicated on the voltmeter.

We see that the effect of the voltmeter has created a change in circuit characteristics and a substantial error in the true voltage

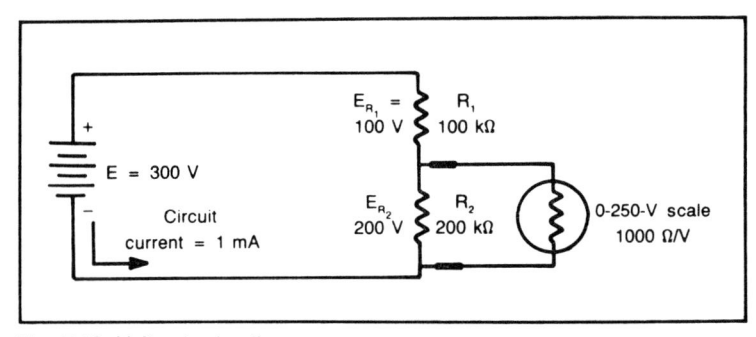

Fig. 7-13. Voltmeter loading.

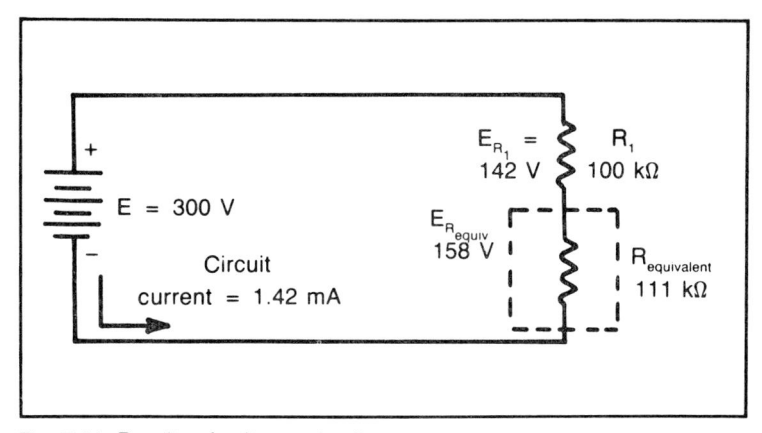

Fig. 7-14. Results of voltmeter loading.

reading. It should be evident then that if a higher-sensitivity meter were connected across the points of measurement the effective resistance of the parallel network would approach the original value.

Digital Voltmeters

All measuring instruments discussed up to this point have been *analog* types. An analog type refers to displaying all measurements with a needle deflection on a continuous scale. In today's digital world, *digital* voltmeters (dvms) have been developed, which display all measurements in discrete numeric readouts.

The advantages of dvms over analog types are

- Human error is minimized
- Reading speed is increased
- Parallax error is eliminated

The heart of a dvm is the electronic circuitry that converts analog voltages to digital form. This conversion process is called analog-to-digital conversion.

THE OHMMETER

An ohmmeter measures the amount of electrical resistance offered by a complete circuit or a circuit component. The proper method of connecting an ammeter in a circuit is illustrated in Fig. 7-15. Notice the very important precaution: *never use an ohmmeter in a circuit until all voltages are removed.*

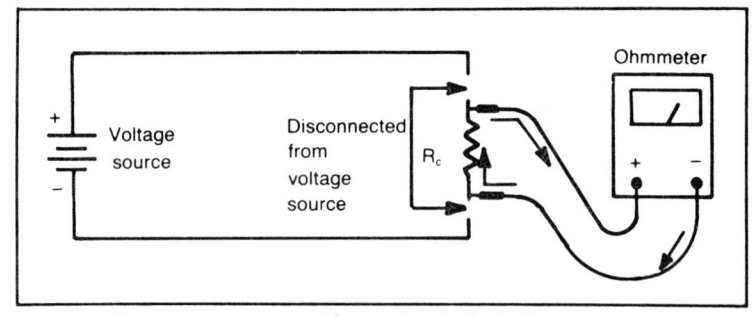

Fig. 7-15. Ohmmeter connected to an electrical circuit.

Ohmmeter Construction

Figure 7-16 shows the principle of an ohmmeter. Note its contained source of emf.

The range of an ohmmeter is from 0 Ω (direct short) to infinity (∞) (open circuit).

Resistor acts as a variable shunt for the meter movement R_a, and these parallel-connected elements are in series with the battery and fixed resistor R_1 which limits the current through the ohmmeter circuit to the amount necessary to deflect the ohmmeter pointer to its full-scale position. When initially adjusting the ohm-

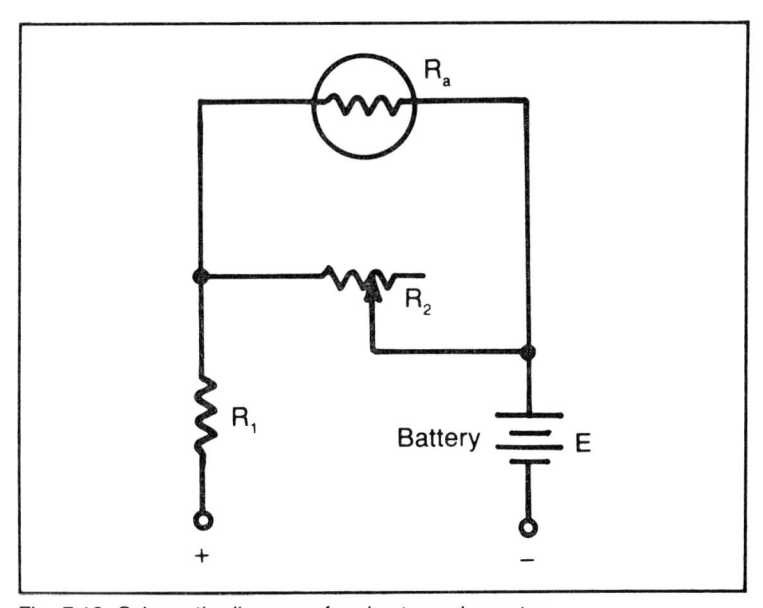

Fig. 7-16. Schematic diagram of series-type ohmmeter.

meter, connect the test leads to the terminals of the ohmmeter and touch the ends together ("short" them) to complete the meter circuit. Adjust variable resistor R_2 so that the meter indicates full-scale deflection or zero. Now that the meter is properly calibrated, put the test leads across the component under test and read the resistance off the meter scale.

An ohmmeter, like the ammeter and voltmeter, usually has more than one range, even though it is calibrated from zero to infinity. Additional ranges are possible by using different values of fixed resistors as meter shunts. Figure 7-17 shows a basic multirange ohmmeter.

Use of Ohmmeters

Although the ohmmeter is one of the most important and versatile test instruments, it is not as accurate as an ammeter or a

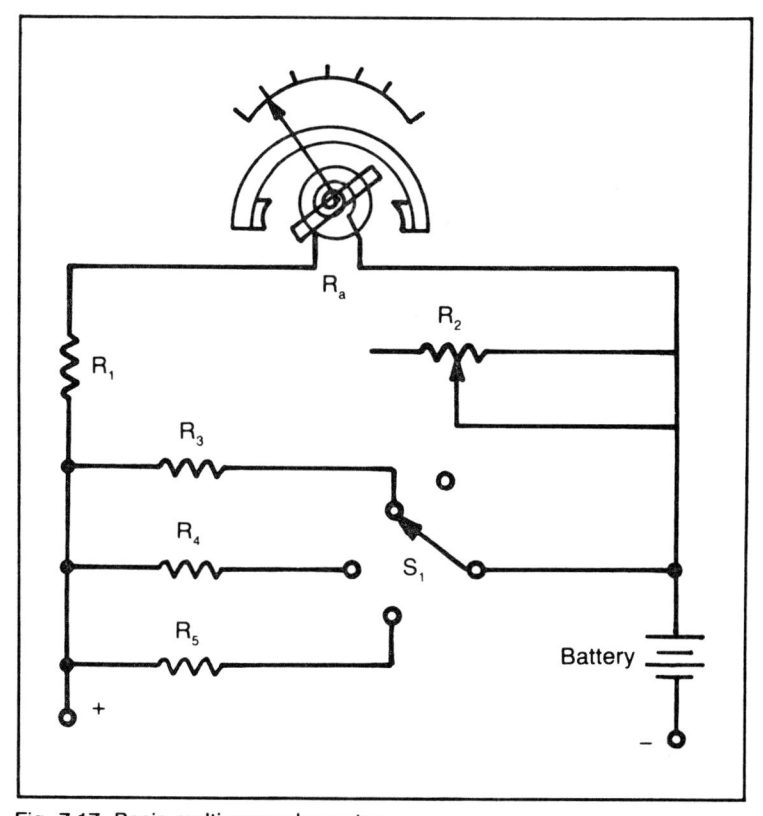

Fig. 7-17. Basic multirange ohmmeter.

131

voltmeter. You should not expect resistance readings with an accuracy of more than 5 or 10 percent.

One of the principal uses of the ohmmeter is checking continuity in a circuit. When troubleshooting electrical circuits or wiring a new circuit, we often make a visual inspection to see if a circuit is complete or if current might be flowing in the wrong part of the circuit because of accidental contact with adjacent components or because of a wiring error. But this is not always the best way. The best method is to send a current through the circuit. If it is complete, current will flow through it. The ohmmeter is perfect for this. It provides the power to send the current through the circuit, and its meter indicates whether current is present. To make such a check, first study the circuit diagram and then check the corresponding parts of the circuit itself with the ohmmeter. The ohmmeter will indicate perfect conduction (zero resistance), partial conduction (when there is resistance), or no conduction (open circuit).

THE MULTIMETER

By combining an ammeter, a voltmeter, and an ohmmeter in a single case, we create a multimeter. This instrument can make almost all of the necessary circuit test measurements in electrical equipment. A multimeter is also referred to as a multitester, multipurpose tester, volt-ohm-ammeter (vom), or circuit analyzer. A key component of a multimeter is the switching arrangement. This allows you to select the proper internal circuits to assure that only one range of one type of measurement is selected at a time.

Construction

Figure 7-18 shows a simplified diagram of a basic multimeter. Note that the three sections of a rotary switch marked S1-A, S1-B, and S1-C are mechanically linked together on a common shaft. This switching arrangement represents the function-selector control. As shown on the diagram, switch positions 1, 2, and 3 represent the ohmmeter, ammeter, and voltmeter positions, respectively.

Care of Meters

Not enough can be said about the care and handling of electrical test meters. They are very sensitive and delicate instruments, and the sudden jarring of a meter can easily damage the fragile construction of the meter movement or bend the meter pointer. The same is true for digital meters.

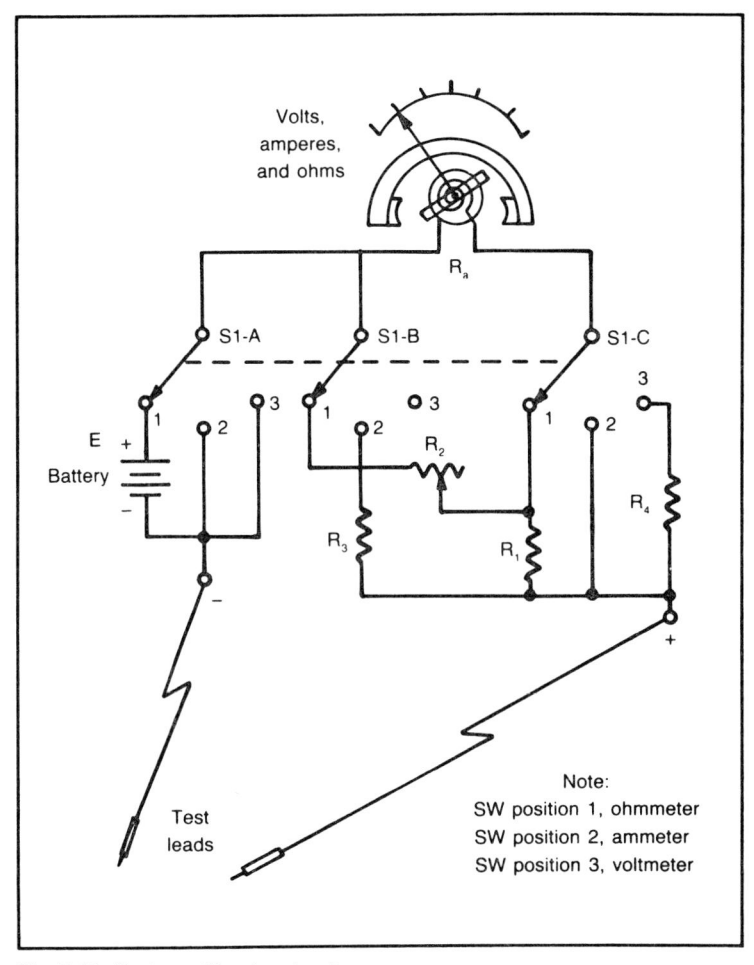

Fig. 7-18. Basic multimeter circuit.

QUIZ

1. An ammeter is always connected in _____ in a circuit.
2. A voltmeter is conected in _____ in a circuit.
3. Voltmeter sensitivity is measured in _____ per _____.
4. Zero ohms of resistance indicates a _____.
5. Infinity indicates an _____.
6. What is parallax error?

133

7. What is a dvm?
8. Does polarity have to be observed when using an ohmmeter?
9. Does polarity have to be observed when using an ammeter?
10. Does polarity have to be observed when using a voltmeter?

Chapter 8

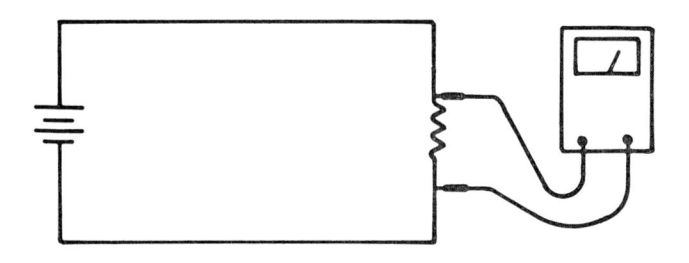

Capacitance

ALTHOUGH CAPACITANCE DOES NOT COME INTO FULL PLAY until it is applied to alternating-current circuits, it does effect dc circuits. This chapter is groundwork for further electrical studies.

THE PROPERTY OF CAPACITANCE

In direct-current circuits voltages are steady and unchanging. However, if a switch is closed or opened, drastic voltage changes can be encountered.

Capacitance opposes any change in voltage. Figure 8-1 illustrates this principle.

Everything exhibits some capacitance. The amount of a capacitance in circuit depends on the physical construction of the circuit and on the electrical devices used. In most cases this capacitance is so small that its effect on voltage is negligible.

Electrical devices that add capacitance to a circuit are called *capacitors* (or *condensers*, although this term is incorrectly used). The circuit symbols for capacitors are shown in Fig. 8-2.

The basic unit of capacitance is the *farad* (F). A capacitor has a capacity of one farad when applied emf of one volt causes the capacitor to take a charge of one coulomb. (One coulomb equals 6.28×10^{18} electrons.) In practical circuits a farad is too large a unit to work with. The *microfarad* ($1\mu F = 10^{-6}$ F) and the

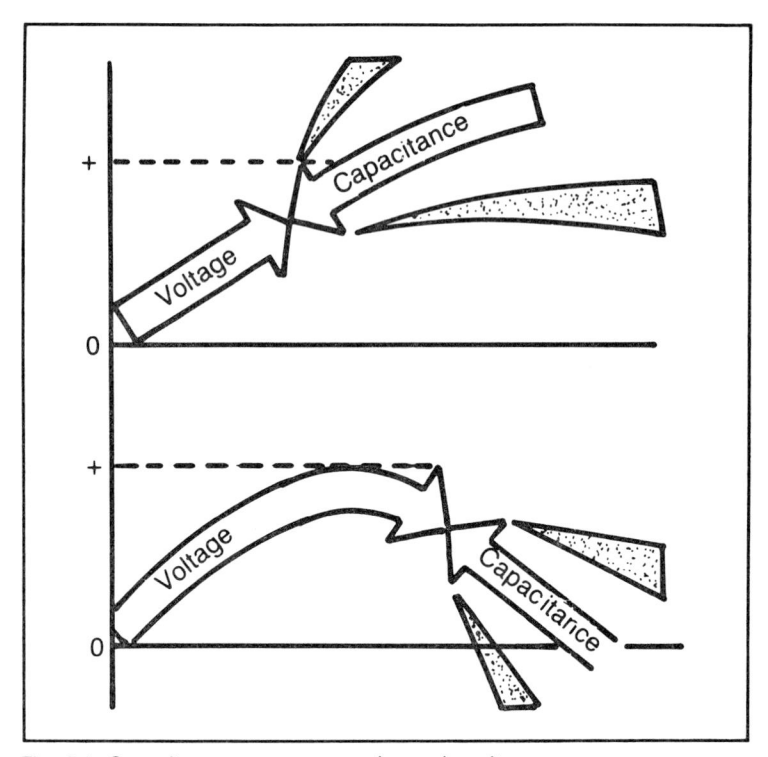

Fig. 8-1. Capacitance opposes any *change* in voltage.

picofarad (1 pF = 10^{-12} F) are the units usually used. See Chapter 3 if you have forgotten how to work with these prefixes.

THEORY OF CAPACITANCE

Capacitance exists because certain parts of a circuit are able to store an electrical charge. Figure 8-3 shows two parallel flat metal plates directly facing each other but not touching. From the principles of static electricity, you can see why these plates can be charged either negatively or positively, depending on the charge

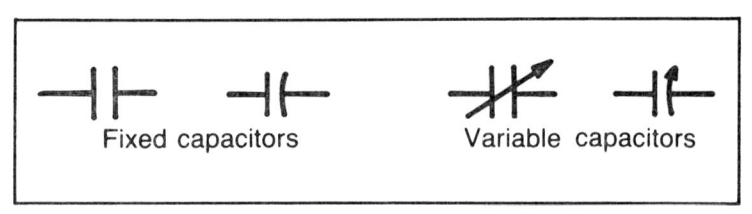

Fig. 8-2. Capacitor symbols.

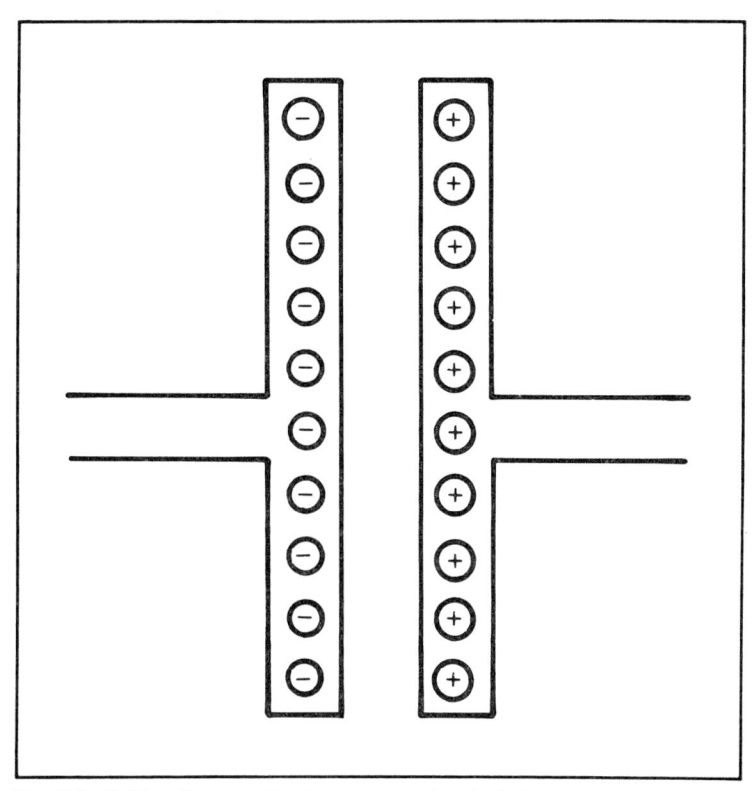

Fig. 8-3. Ability of a capacitor to store an electrical charge.

transferred to them. The positively charged plate will be forced to give up some of its electrons, and the negatively charged plate will have an excess of electrons pumped into it.

To charge the plates, we need an emf. For the plate to have negative charge, extra electrons from a negative source of emf must be forced onto the plate. The first few extra electrons go onto the plate quite readily, but once there they tend to oppose or repel any additional electrons that try to follow. As more and more of these extra electrons are forced onto the plate, the repelling force increases so that a greater emf is needed to keep up the movement. When this repelling force is equal to the charging force, all movement stops because no more electrons can be forced onto the plate. Figure 8-4 illustrates this action.

Likewise, if electrons are removed from a plate by the attraction of a positive charge, the plate is positively charged. Figure 8-5 illustrates this action. The first few electrons leave quite readily,

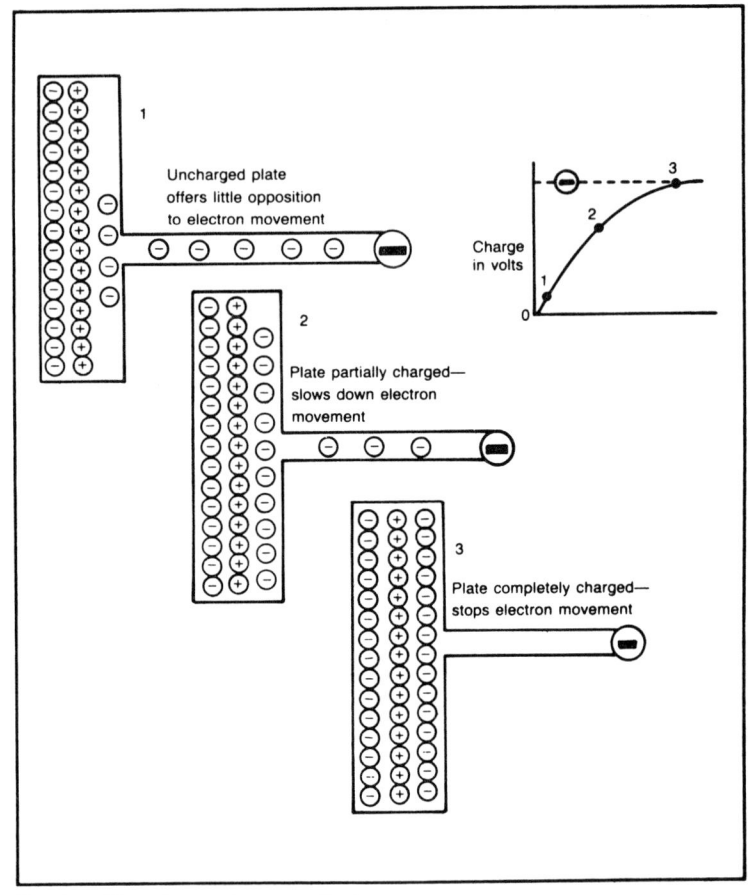

Fig. 8-4. Charging a plate negatively.

but as more electrons move, a stronger positive charge is built up on the plate. This positive charge attracts electrons and makes it more difficult to pull them away. When this positive attracting force equals the charging force, all movement stops and no more electrons leave the plate.

CAPACITIVE TIME CONSTANT

If the plates of the capacitor are uncharged and the switch is open, no current flows and no voltage exists between the plates of the capacitor (see Fig. 8-6). When the switch is closed, the battery furnishes electrons to the plate of the capacitor connected to the negative terminal of the battery. At the same time, electrons are taken away from the plate of the capacitor connected to the positive terminal of the battery.

At the moment the switch is closed, the voltage across the plates of the capacitor should equal the battery voltage of 3.0 V; however, this does not occur instantly because one plate of the capacitor must have time to accumulate excessive electrons to obtain a negative charge, and the other plate must have time to give up an equal number of electrons to become positively charged. This

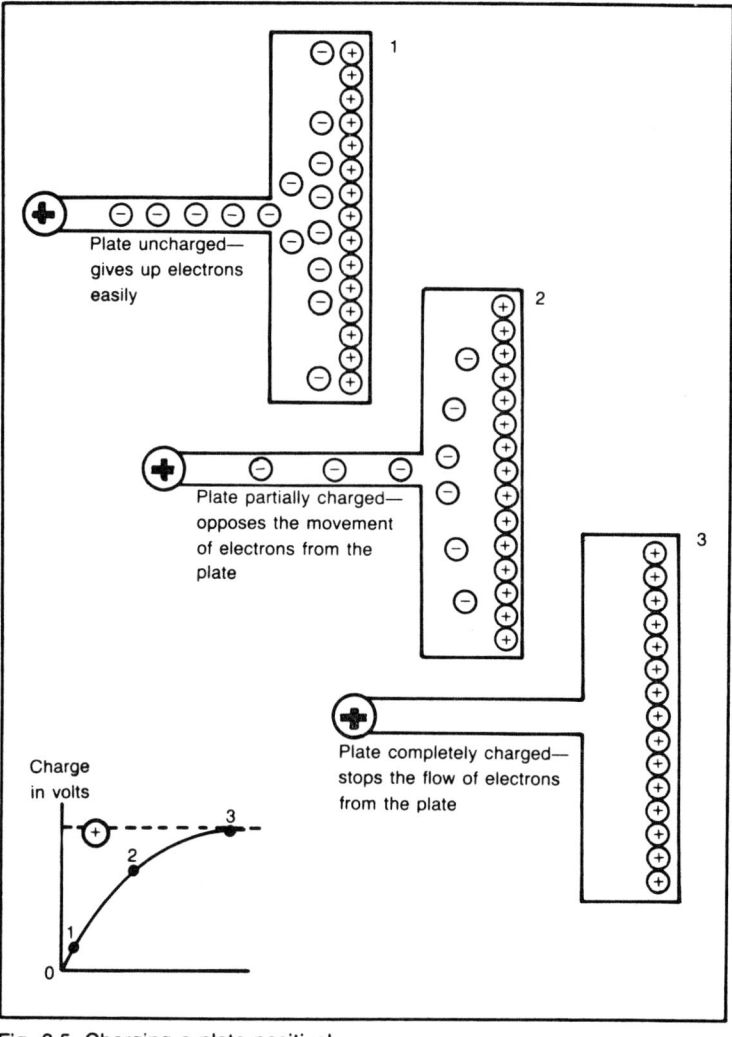

Fig. 8-5. Charging a plate positively.

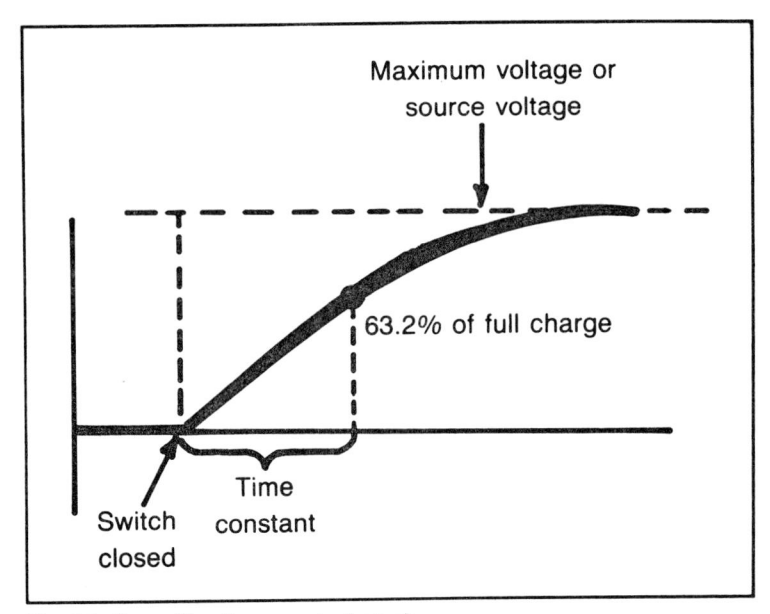

Fig. 8-6. Capacitive time constant graph.

action on the two plates of a capacitor in a dc circuit is called *capacitance*. Capacitance opposes the change in voltage from 0 V, with the switch in the off position, to 3.0 V, with the switch in the closed position. It delays this change for a limited amount of time, but it does not prevent it.

When the switch is opened, the capacitor remains fully charged because there is no path for these electrons to follow. At this point, if the switch is again closed, the voltage across the plates of the capacitor will *instantaneously* be 3.0 V, and there is no capacitive effect on the circuit because the capacitor was already charged.

You must remember two main points about capacitance in *dc circuits*:

- When the capacitor is fully charged, no further current flows.
- A period of time is required to charge a capacitor.
- The time required for a capacitor to become charged depends on the amounts of resistance and capacitance in the circuit. This time is known as the *time constant*. A simple formula for finding the time constant is

$$T_c = R \times C$$

where: T_C = time constant in seconds
R = resistance in ohms
C = capacitance in farads

This time constant is the time in seconds required for the voltage across a capacitor to reach 63.2 percent of its maximum value. Note the graph shown in Fig. 8-6C.

The time required for the capacitor to reach its full charge is five times the amount of time it took for it to reach 63.2 percent of its charge. Therefore,

$$T_{fc} = T_c \times 5$$

where: T_{fc} = time in seconds for the capacitor to reach its full charge
T_c = time constant in seconds

FACTORS AFFECTING CAPACITANCE

A capacitor consists of two plates that can be charged and

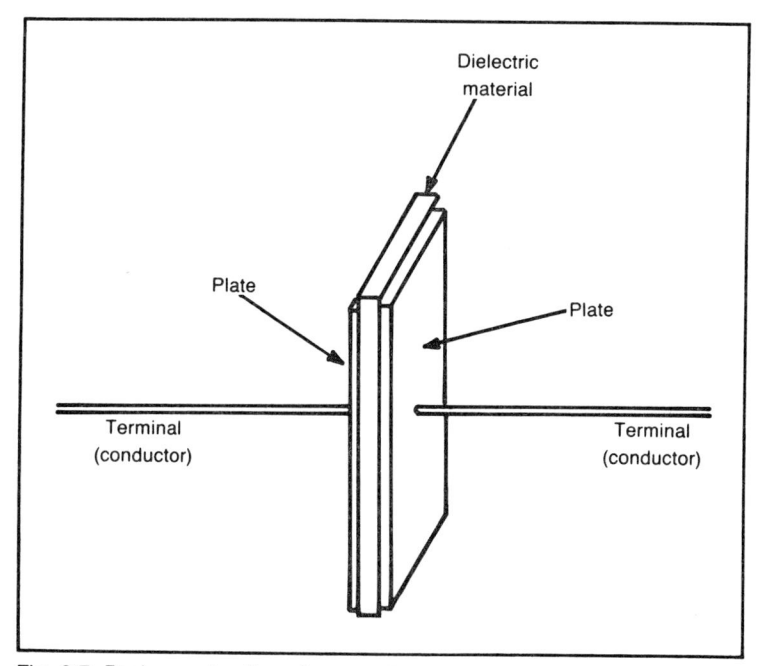

Fig. 8-7. Basic construction of a capacitor.

separated by an insulating material called the *dielectric*. Early capacitors were made with solid metal plates; newer types of capacitors use metal foil, especially aluminum, for the plates. Dielectric materials commonly used include air, mica, and waxed paper. Figure 8-7 illustrates the basic construction of a capacitor.

Three physical factors influence the capacity of a capacitor:

- The area of the plates
- The distance between the plates (thickness of the dielectric material)
- The material used for the dielectric

Plate Area

Capacitance varies directly with the cross-sectional area of the plates. A large plate area has room for more electrons than a small plate area; therefore, it can hold a greater charge. Likewise, a larger plate area has more electrons to give up and will hold a much larger positive charge than a smaller plate area. Therefore, an increase in plate area increases capacitance, and a decrease in plate area decreases capacitance. Figure 8-8 illustrates this fact.

Plate Spacing

The capacitance between two plates increases as the plates are brought closer together and decreases as the two plates are moved apart. This occurs because the charge on one plate has a stronger effect on the charge of the other plate as the plates are moved nearer to each other; the effect decreases as the plates are moved apart. Remember, that when an excess of electrons appears on one plate of a capacitor, electrons are forced off the opposite plate, inducing a positive charge in this plate. Likewise, a positively charged plate induces a negative charge on the opposite plate. The closer the plates are to each other, the stronger the force between them. This force increases the capacitance of the circuit. Figure 8-9 illustrates this.

Dielectric Material

The capacitance changes if a different insulating material is used for the dielectric. The effect of different materials is compared to that of air. Materials other than air will multiply the capacitance by a certain amount, called the *dielectric constant*. Some types of waxed paper have a dielectric constant of 3. If this waxed paper

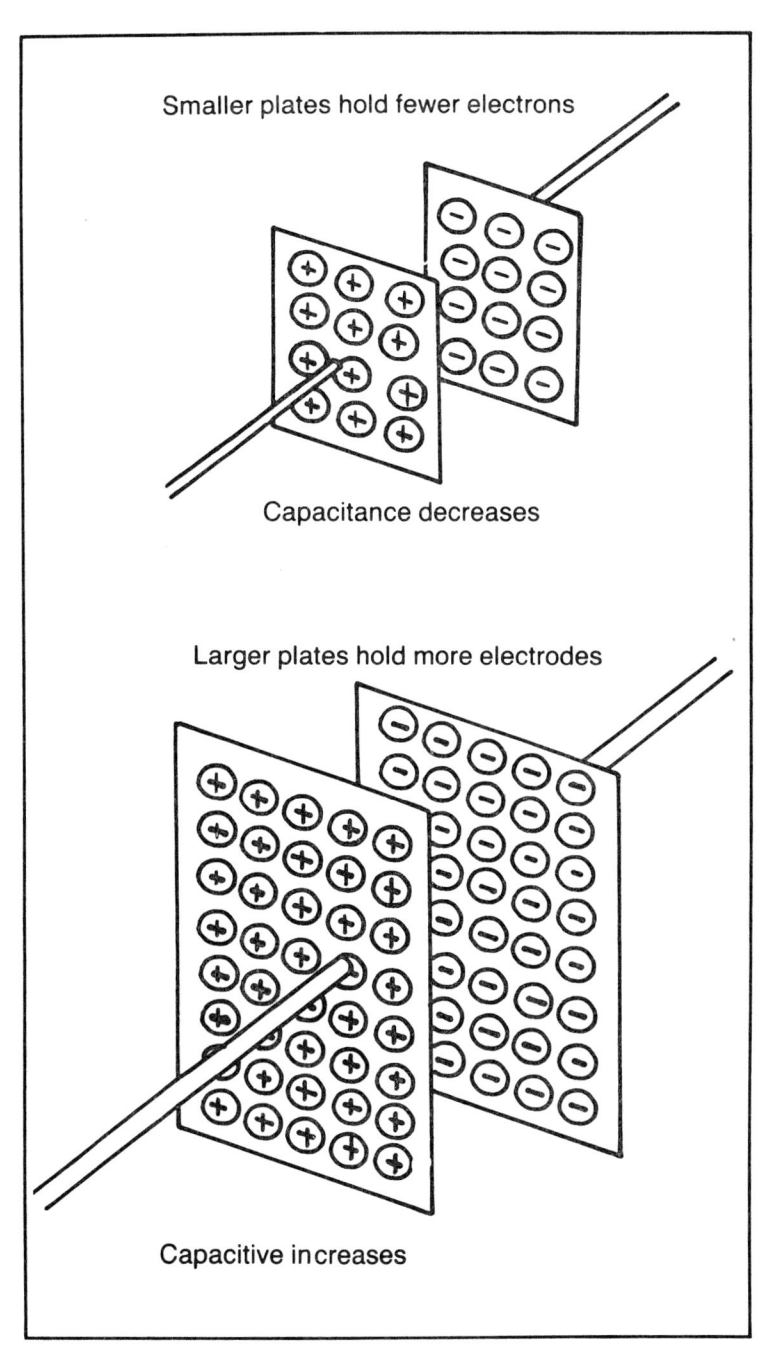

Fig. 8-8. Effect of increasing plate area.

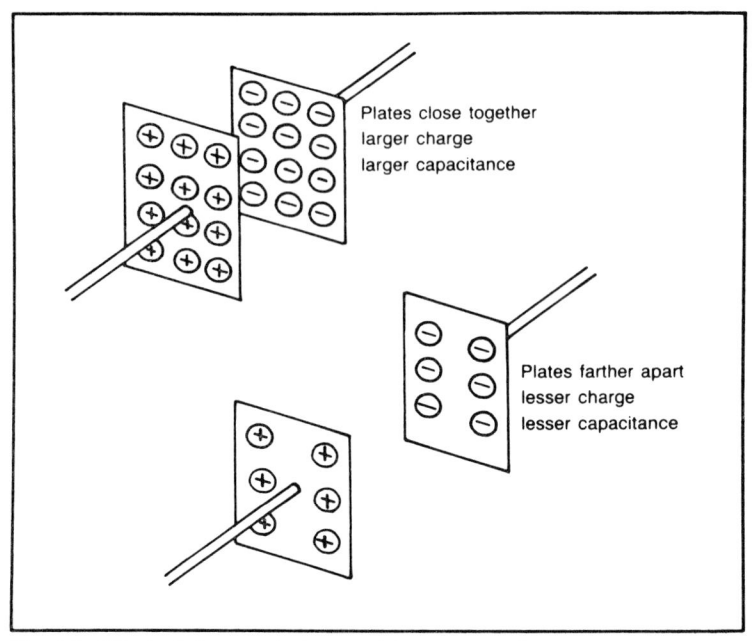

Fig. 8-9. Effect of distance between plates on capacitance.

is placed between the plates, the capacitance will be three times greater than the capacitance if air were used as the dielectric. Table 8-1 lists different dielectric constants.

CALCULATING TOTAL CAPACITANCE

Connecting capacitors in series decreases total capacitance because it effectively increases the distance between the plates.

Table 8-1. Dielectric Constants of Insulators.

Material	Constant (k)
Air	1
Bakelite	6.6 to 16
Cloth (varnished)	3.5 to 5.5
Glass (hard)	5.5 to 10
Mica	3 to 6
Paper (impregnated with oil or wax)	2.5 to 4
Quartz	4.3 to 5.1
Rubber	2.5 to 3
Slate	6 to 7
Sulfur	4

Figure 8-10A illustrates this fact. To find the total capacitance of series-connected capacitors, use the following formulas (note the similarity to calculating total resistance for parallel-connected resistors):

$$C_t = \frac{\text{Capacitance of 1 capacitor}}{\text{No. of capacitors in the circuit}} \quad \text{(equal capacitance method)}$$

$$C_t = \frac{1}{\frac{1}{C_1} + \frac{1}{C_2} + \frac{1}{C_3} + \ldots} \quad \text{(reciprocal method)}$$

$$C_t = \frac{C_1 \times C_2}{C_1 + C_2} \quad \text{(product over sum method)}$$

When capacitors are connected in parallel, total capacitance increases because the total plate area has increased. Figure 8-10B illustrates this fact. The total capacitance for parallel-connected

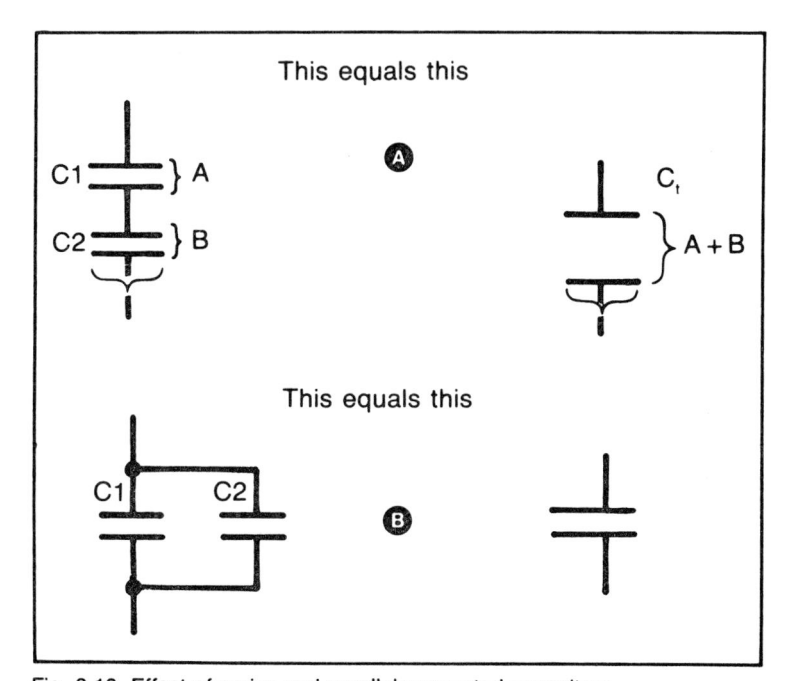

Fig. 8-10. Effect of series and parallel-connected capacitors.

capacitors is found by simply adding the individual capacitances (just like finding total resistance in a series circuit):

$$C_t = C_1 + C_2 + C_3 + \ldots$$

TYPES OF CAPACITORS

Many types of capacitors are used in electricity and electronics. Capacitors are generally classified according to the dielectrics. The most common types are air, mica, paper, and ceramic.

Air Capacitors

The air capacitor is the most basic type. It is constructed of metal plates with an air space between the plates. The capacitance

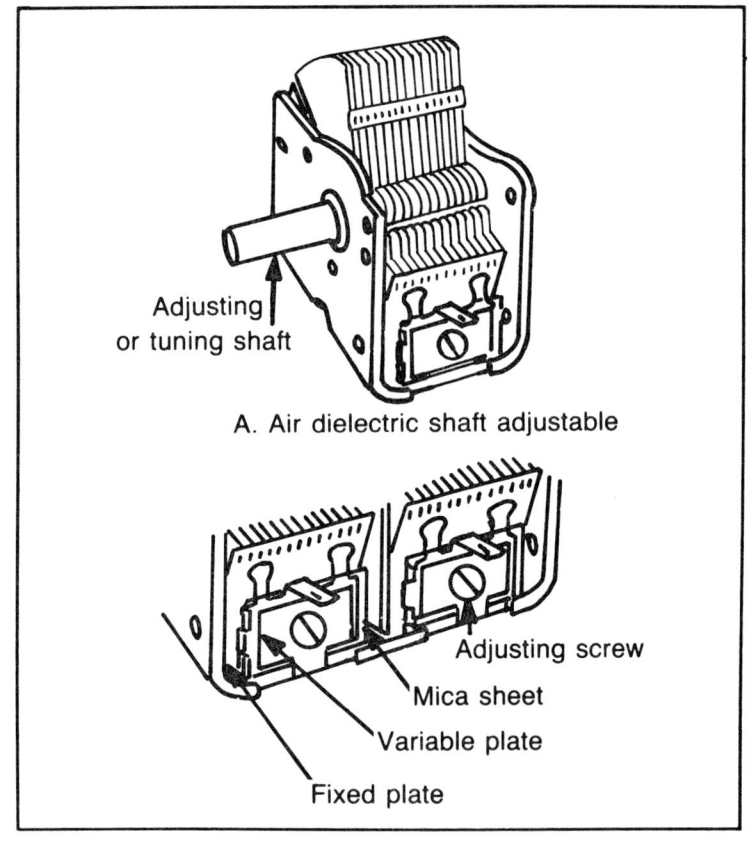

Fig. 8-11. Air capacitors.

Fig. 8-12. Mica capacitors.

of air capacitors ranges from 1 to 500 pF. Figure 8-11 shows the construction of different types of air capacitors.

Mica Capacitors

Mica capacitors are made of thin metal-foil plates separated by a sheet of mica and molded into a plastic package. These capacitors have a range from 10 pF to 0.01 μF. Different types of terminals are used to enable these capacitors to be placed into circuits. Figure 8-12 shows this type of construction.

The voltage applied across the plates of a capacitor is determined by the thickness of the dielectric. We have already found out that if we increase the thickness of the dielectric, the distance between the plates is effectively increased and the capacitance is decreased. To compensate for this, we must increase the size of the plates. Therefore, capacitors with high voltage ratings are usually large. Figure 8-13 illustrates these comparisons.

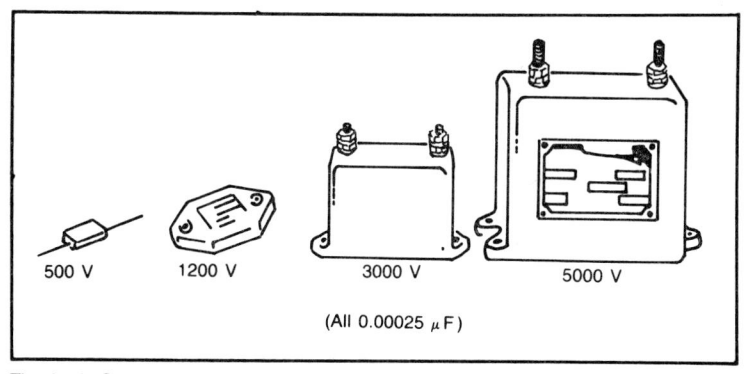

500 V   1200 V   3000 V   5000 V

(All 0.00025 μF)

Fig. 8-13. Comparisons of physical size of capacitors of different voltage ratings.

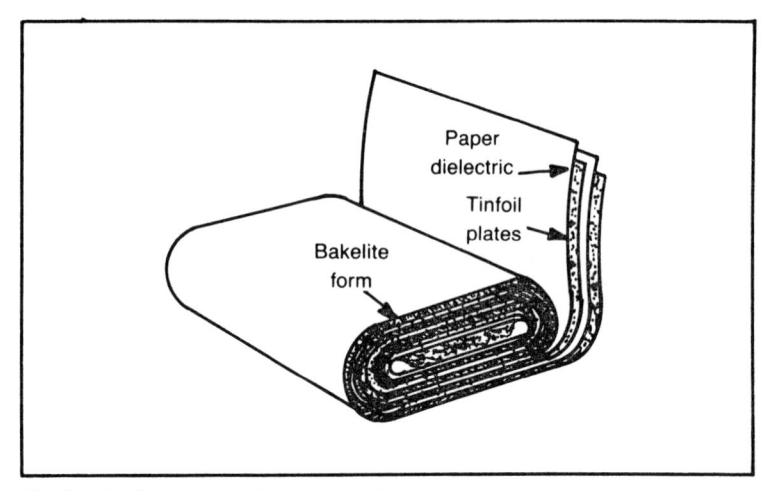

Fig. 8-14A. Paper capacitor construction.

Paper Capacitors

Paper capacitors are constructed of strips of metal foil separated by strips of waxed paper. The range of paper capacitors is from 250 pF to 1 μF. Because very long strips of paper are required to get a usable value of capacitance, the strips of foil and waxed paper are rolled together to form a cartridge. This package, including the leads, is sealed in wax to prevent leakage and corrosion of the plates. Figure 8-14 shows examples of this process and some examples of what paper capacitors look like.

Ceramic Capacitors

Ceramic capacitors are extremely small; they use ceramic as the dielectric and a silver-film deposit for the plates. They cover a range from 1 pF to 0.02 μF and are suitable for very high-voltage

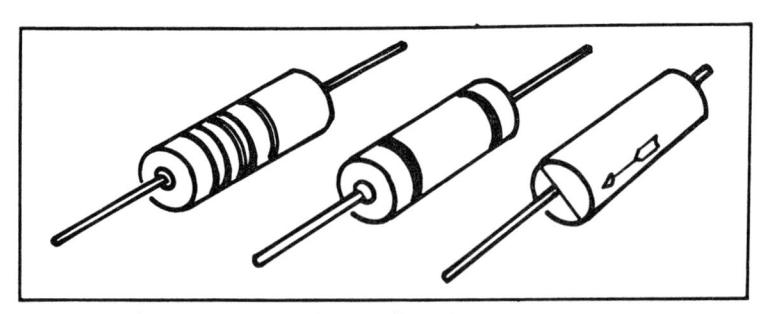

Fig. 8-14B. Some paper capacitor configurations.

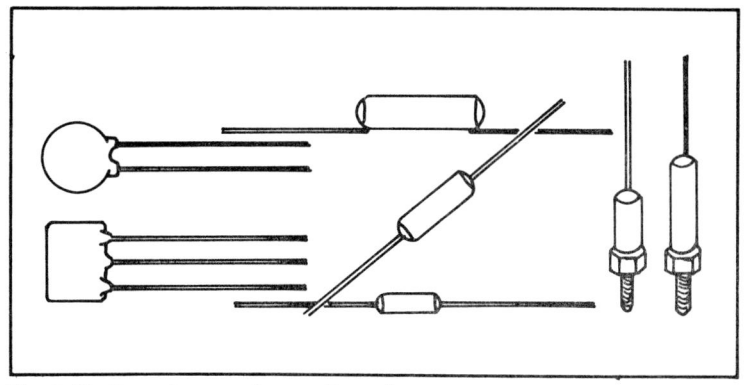

Fig. 8-15. Ceramic capacitor configurations.

applications. Figure 8-15 shows some ceramic capacitor configurations.

SPECIAL TYPES OF CAPACITORS

Bathtub Capacitors

Bathtub capacitors are paper capacitors that have been hermetically sealed in a metal container. This metal container sometimes acts as one of the terminals; at other times it acts as a shield against any electrical interference.

Metal-Encased Capacitors

Capacitors used in automobile ignition systems are also metal-encased paper capacitors. Metal cases are necessary because automobile capacitors must be exceptionally rugged and able to withstand both weather and mechanical shock. Figure 8-16 shows bathtub and metal-encased capacitors.

Electrolytic Capacitors

For capacitances greater than 1 μF, the physical size of paper or mica capacitors becomes too large. Electrolytic capacitors are used for ranges from 1 to 1000 μF. Unlike other capacitors, an electrolytic capacitor is polarized with definite negative and positive plates. If improperly connected in a circuit, it will break down and act as a direct short. Electrolytic capacitors are constructed in a wide variety of shapes and sizes, as shown in Fig. 8-17.

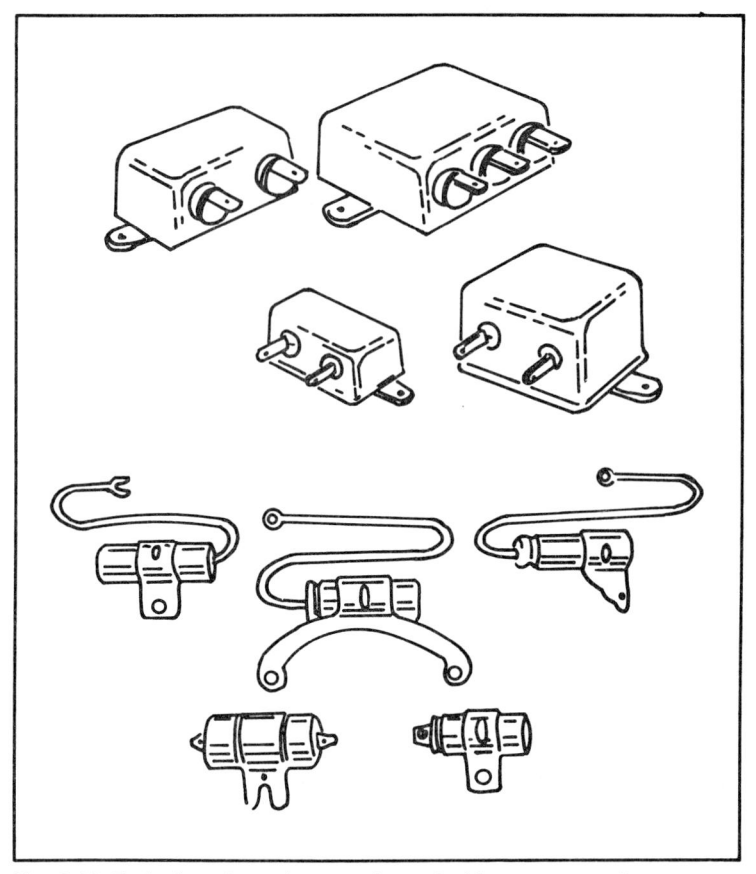

Fig. 8-16. Bathtub and metal encased auto ignition type capacitors.

COLOR CODE

Most capacitors have their capacitance values and voltage ratings stamped on their bodies. Electrolytic capacitors also have polarity markings. A voltage rating stamped on a capacitor refers to the maximum dc voltage that can be applied across the terminals without breaking down the dielectric.

In many cases, capacitors are marked with a color code quite similar to the resistor color code; in fact, the corresponding colors and numbers are the same for both codes. With few variations, this basic color code is generally the same for all types of capacitors. All variations express the values of capacitance in picofarads. Figure 8-18 shows what is called the dot color-code system for Electronic Industries Association (EIA) mica and molded paper capacitors. If the three dots in the top row were white, orange, and blue, from left to right, and the multiplier dot were orange, the value of the capacitor in picofarads would be determined as follows:

1. White in the type dot indicates a mica capacitor.
2. Orange and blue indicate the two-digit number 36.
3. The orange multiplier indicates that three zeros must be added to the two-digit number to obtain the capacitance in picofarads.

Hence, the value of the capacitance is 36,000 pF or 0.036 μF.

The two dots remaining in this system represent the tolerance and the characteristic. The tolerance dot represents any value from 1 to 10 percent, 20 percent if the dot is the same color as the body of the capacitor. The sixth dot, rarely used, pertains to the temperature coefficients and methods of testing.

A five-dot marking system for molded paper and mica capacitors is shown in Figure 8-19. Note that the reading direction

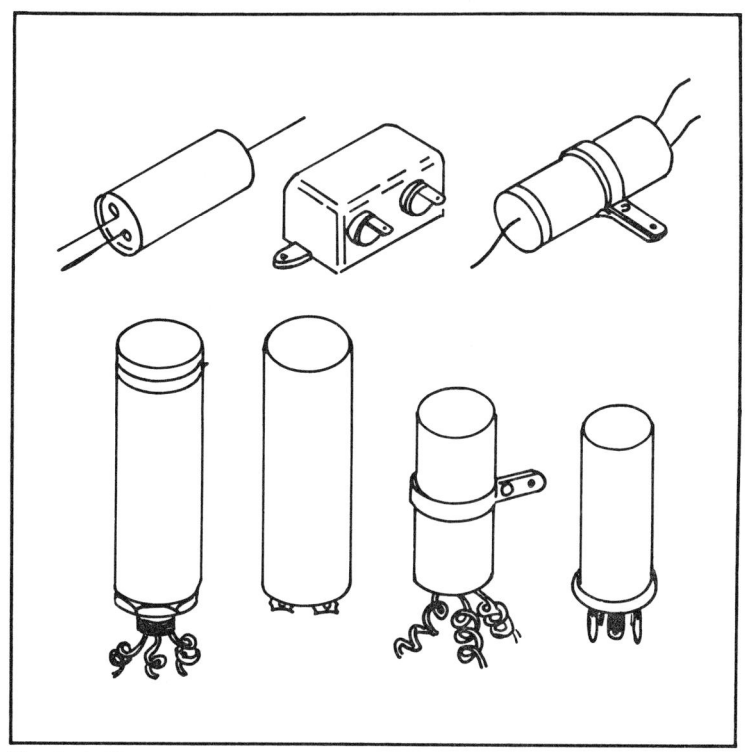

Fig. 8-17. Electrolytic capacitor configurations.

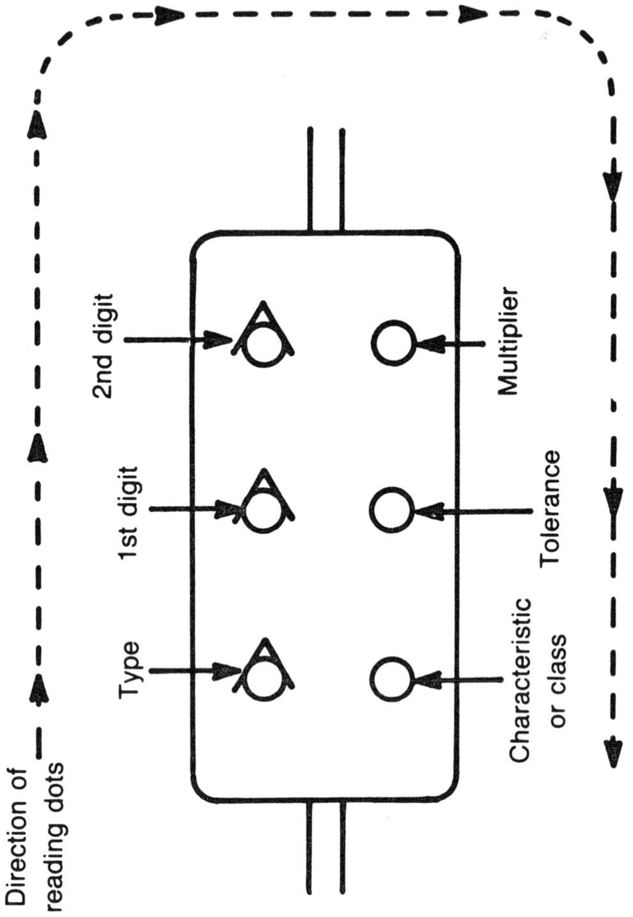

Direction of reading dots
Type
1st digit
2nd digit
Multiplier
Tolerance
Characteristic or class

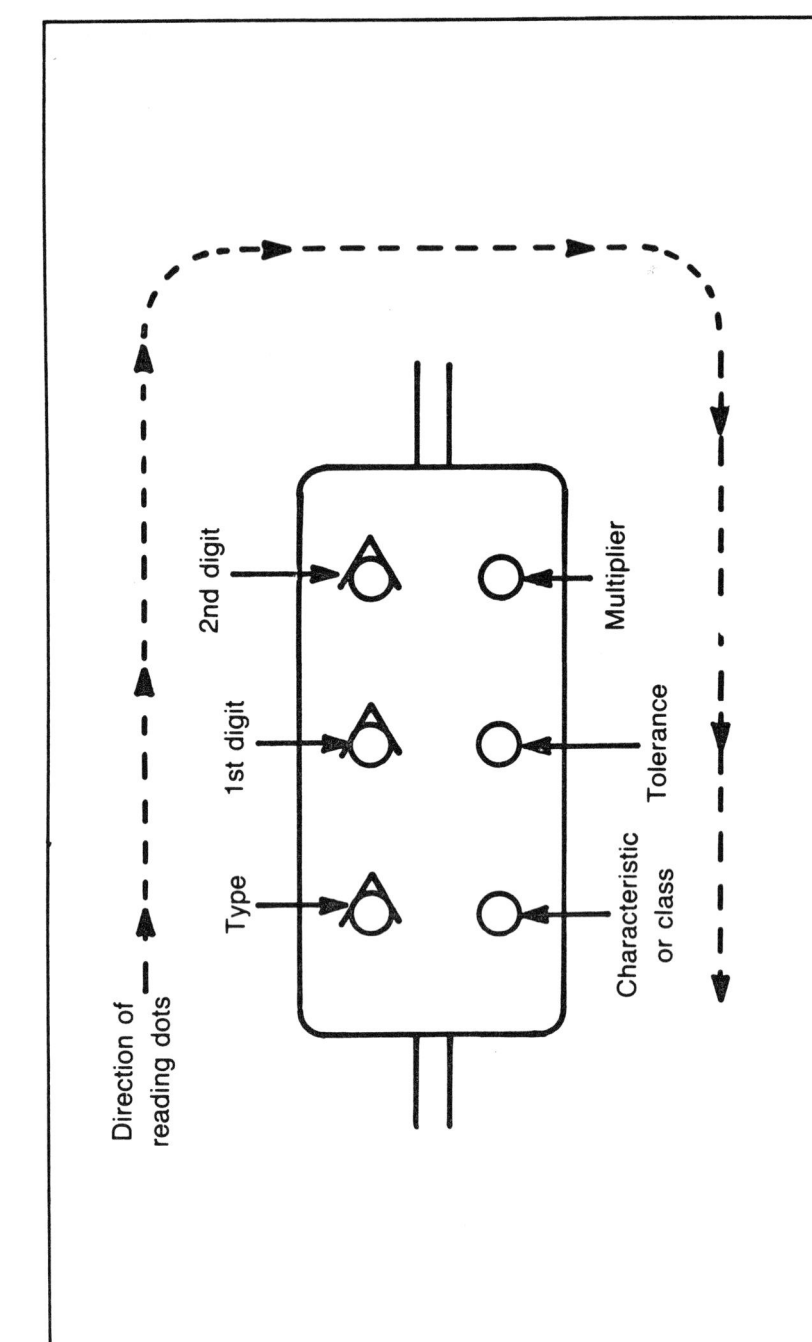

Type	Color	1st digit	2nd digit	Multiplier	Tolerance (percent)	Characteristic or class
Mica	Black	0	0	1.0		Applies to temperature coefficient or methods of testing
	Brown	1	1	10	± 1	
	Red	2	2	100	± 2	
	Orange	3	3	1,000	± 3	
	Yellow	4	4	10,000	± 4	
	Green	5	5	100,000	± 5	
	Blue	6	6	1,000,000	± 6	
	Violet	7	7	10,000,000	± 7	
	Gray	8	8	100,000,000	± 8	
EIA, Mica	White	9	9	1,000,000,000	± 9	
	Gold			0.1		
Molded paper	Silver			0.01	±10	
	Body				±20	

Fig. 8-18. Six-dot color code system for EIA mica and molded paper capacitors.

154

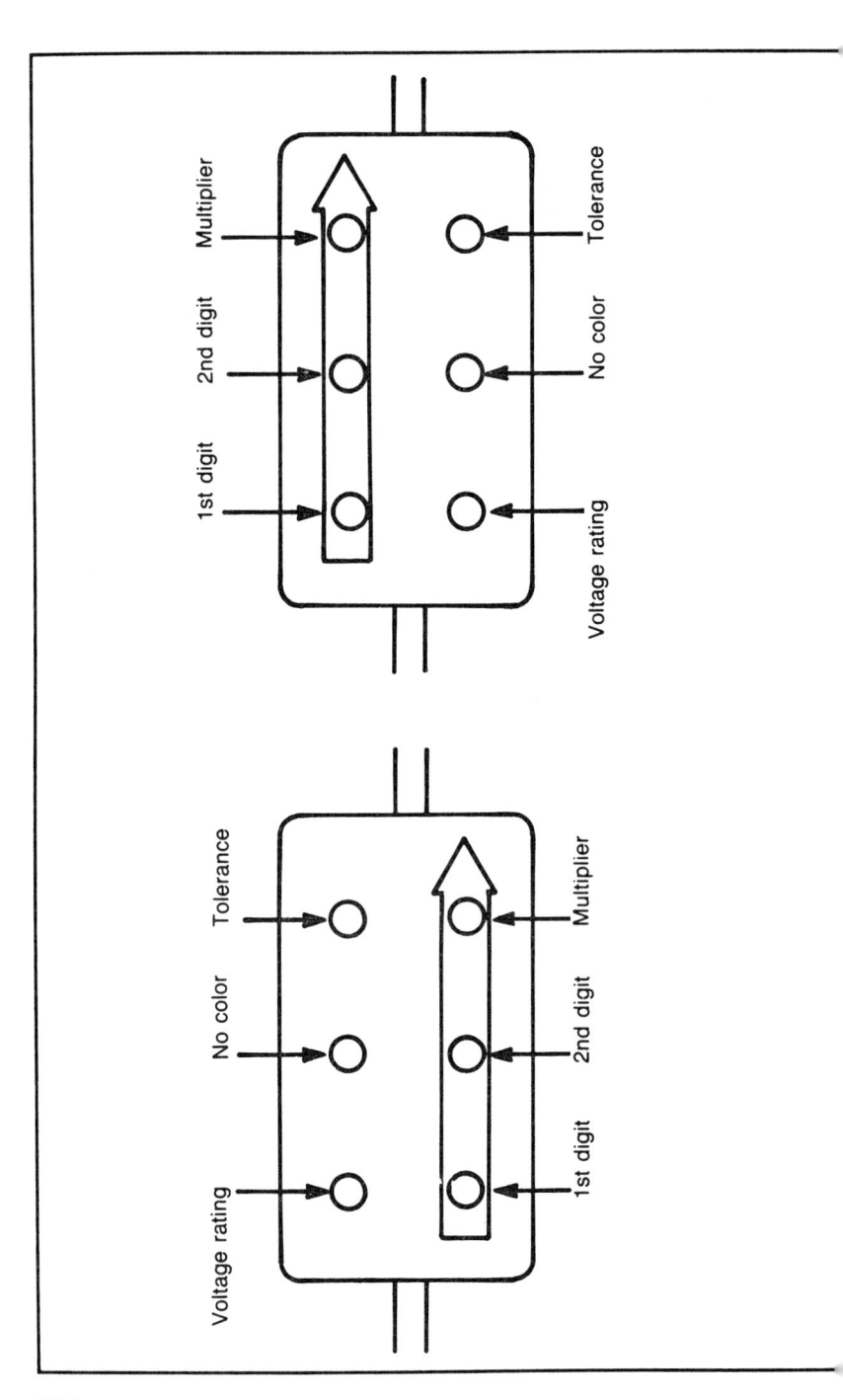

Color	1st digit	2nd digit	Multiplier	Tolerance (percent)	Voltage rating
Black	0	0	1.0	± 1	100
Brown	1	1	10	± 2	200
Red	2	2	100	± 2	200
Orange	3	3	1,000	± 3	300
Yellow	4	4	10,000	± 4	400
Green	5	5	100,000	± 5	500
Blue	6	6	1,000,000	± 6	600
Violet	7	7	10,000,000	± 7	700
Gray	8	8	100,000,000	± 8	800
White	9	9	1,000,000,000	± 9	900
Gold			0.1		1000
Silver			0.01	±10	2000
Body				±20	*

*Where no color is indicated, the voltage rating may be as low as 300 V

Fig. 8-19. Five-dot capacitor color-code system.

is different from the six-dot system. In the five-dot system the first digit is the capacitor value. Suppose the capacitor is marked green, violet, and orange. It would be read as follows:

$$\underset{5}{\text{1st Digit}} \quad \underset{7}{\text{2nd Digit}} \quad \times \quad \underset{1000}{\text{Multiplier}} = 57{,}000 \text{ pF}$$

The final two dots, representing tolerance and voltage rating, are always read from right to left. If this capacitor had a white tolerance dot, its true value would be within ±9 percent of the indicated value. If the color of the last dot is the same as the body color of the capacitor, you should assume that the maximum rating of the capacitor is 300 V. Never exceed the ratings marked on the capacitor. Serious damage could occur.

The color-band system for coding tubular paper capacitors uses six color bands for marking the capacitance, tolerance, and voltage rating. An example is shown in Fig. 8-20. This figure shows that the first four bands, which begin near one end of the capacitor, relate to capacitance. You should turn the capacitor so that the bands may be read from left to right as shown. The main difference between this system and the others is that two stripes are used to indicate voltage rating. The voltage rating is found by taking the coded values and multiplying by 100. For example, if this capacitor has orange, blue, brown, black, brown, black color bands, it would be a 36.0 pF ±20 percent with a voltage rating of 1000 V. If the capacitor's working voltage is less than 1000 V, the second voltage band will be omitted.

QUIZ

1. Capacitance is the property that opposes any change in

    A. Voltage
    B. Current
    C. Resistance

2. The capacitive time constant is the time in seconds required for the voltage across a capacitor to reach _____ percent of its maximum value.
3. 0.00450 μF = _____ pF?
4. Name the three physical factors that affect capacitance.
5. The value of C_t in Fig. 8-21 is _____.

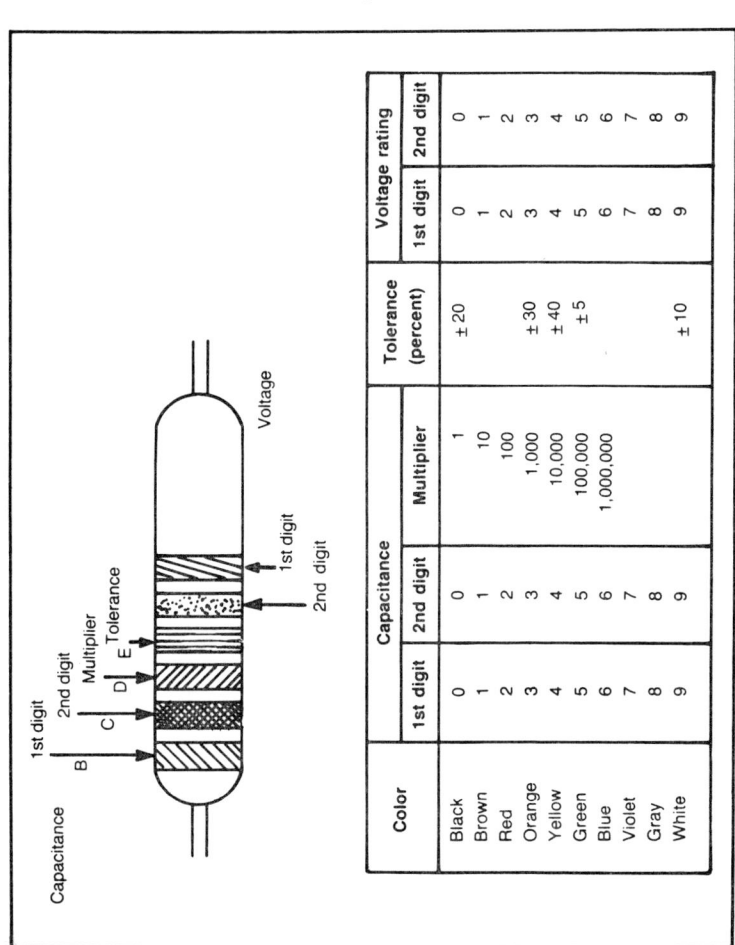

| Color | Capacitance | | | Tolerance (percent) | Voltage rating | |
	1st digit	2nd digit	Multiplier		1st digit	2nd digit
Black	0	0	1	±20	0	0
Brown	1	1	10		1	1
Red	2	2	100		2	2
Orange	3	3	1,000	±30	3	3
Yellow	4	4	10,000	±40	4	4
Green	5	5	100,000	±5	5	5
Blue	6	6	1,000,000		6	6
Violet	7	7			7	7
Gray	8	8			8	8
White	9	9		±10	9	9

Fig. 8-20. Tubular paper capacitor color-code system.

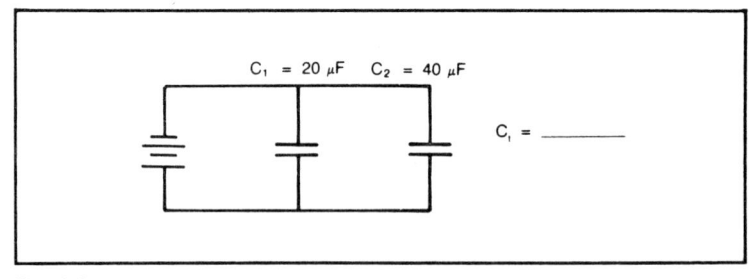

Fig. 8-21

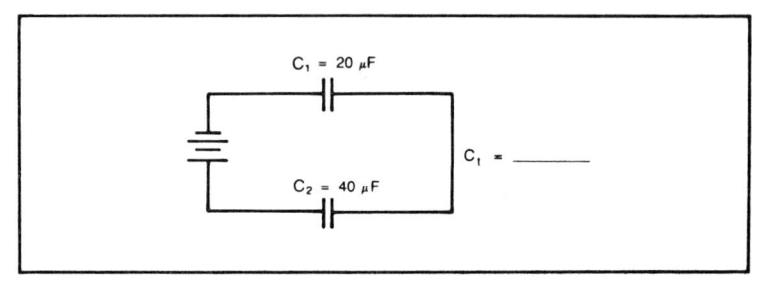

$C_1 = 20\ \mu F$ $C_2 = 40\ \mu F$

$C_t = $ _____

Fig. 8-22.

$C_1 = 20\ \mu F$

$C_2 = 40\ \mu F$

$C_t = $ _____

6. The value of C_t in Fig. 8-22 is _____.
7. Convert 550 μF to F.
8. How many time constants are required before the voltage across a capacitor reaches its full value?
9. Which has the higher dielectric constant: air or mica?
10. The basic unit of capacitance is the _____.

Chapter 9

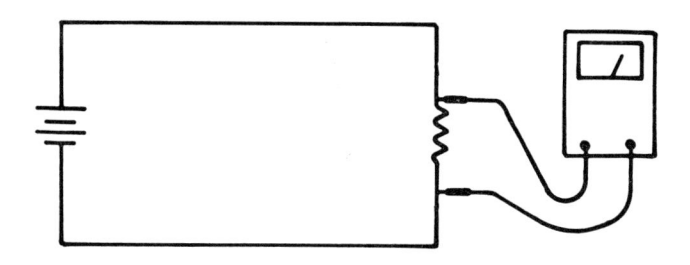

Inductance

INDUCTANCE IS THE PROPERTY TENDING TO OPPOSE ANY change in current. Inductance opposes both momentary increases and decreases in current. It is also a measure of the ability of an electric circuit to store and release magnetic energy.

Although inductance is primarily an ac phenomenon, every electrical circuit contains some inductive properties, so inductance must always be considered. Devices that purposely add inductance to a circuit are called *inductors*. Inductors come in many shapes and sizes, but they are basically nothing more than a coil of wire. In practical applications, names such as choke, choke coil, inductance coil, inductive reactor, and reactor are used to indicate inductors. Some inductors are shown in Figs. 9-1, 9-2, and 9-3.

SELF-INDUCTANCE

When current starts to flow through a straight current-carrying conductor, a magnetic field builds up. This field surrounds the conductor in concentric circles and increases in intensity as the current increases. During this time of increase, the flux is also increasing. A voltage is induced within the conductor that sends a current in the direction *opposite* to the increasing current. This induced voltage is what tends to oppose the change in current.

Self-induction is shown in Fig. 9-4. When the current in one turn of the coil suddenly increases, the lines of force (flux) around that turn expand and move outward. As they do so, they cut over

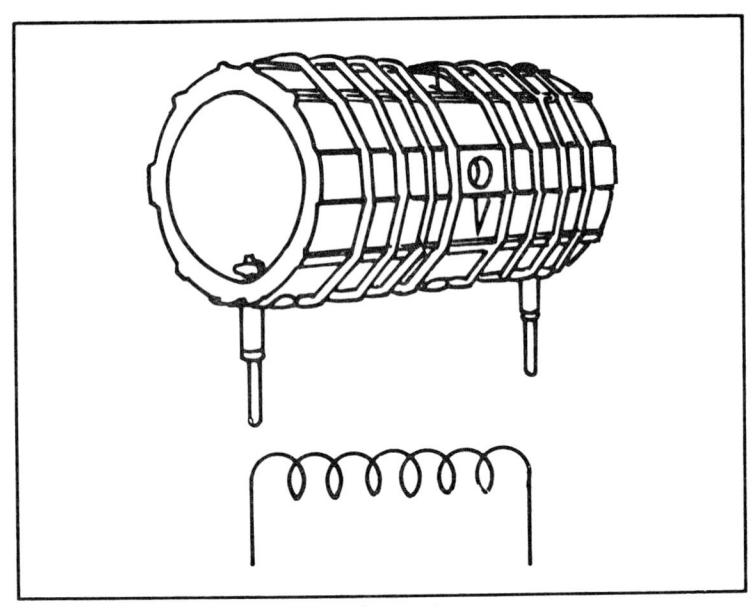

Fig. 9-1. Air core fixed inductor and symbol.

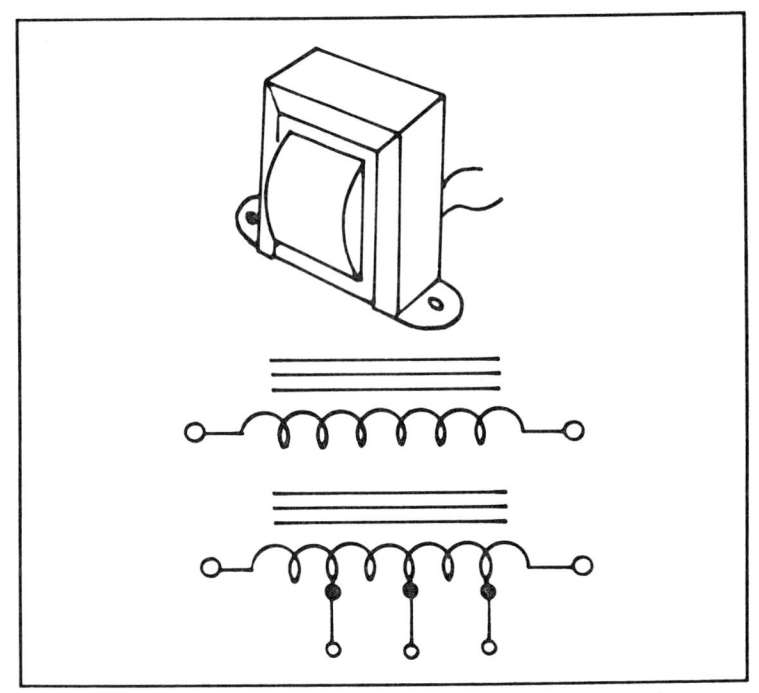

Fig. 9-2. Iron core inductor and symbols for both fixed and tapped types.

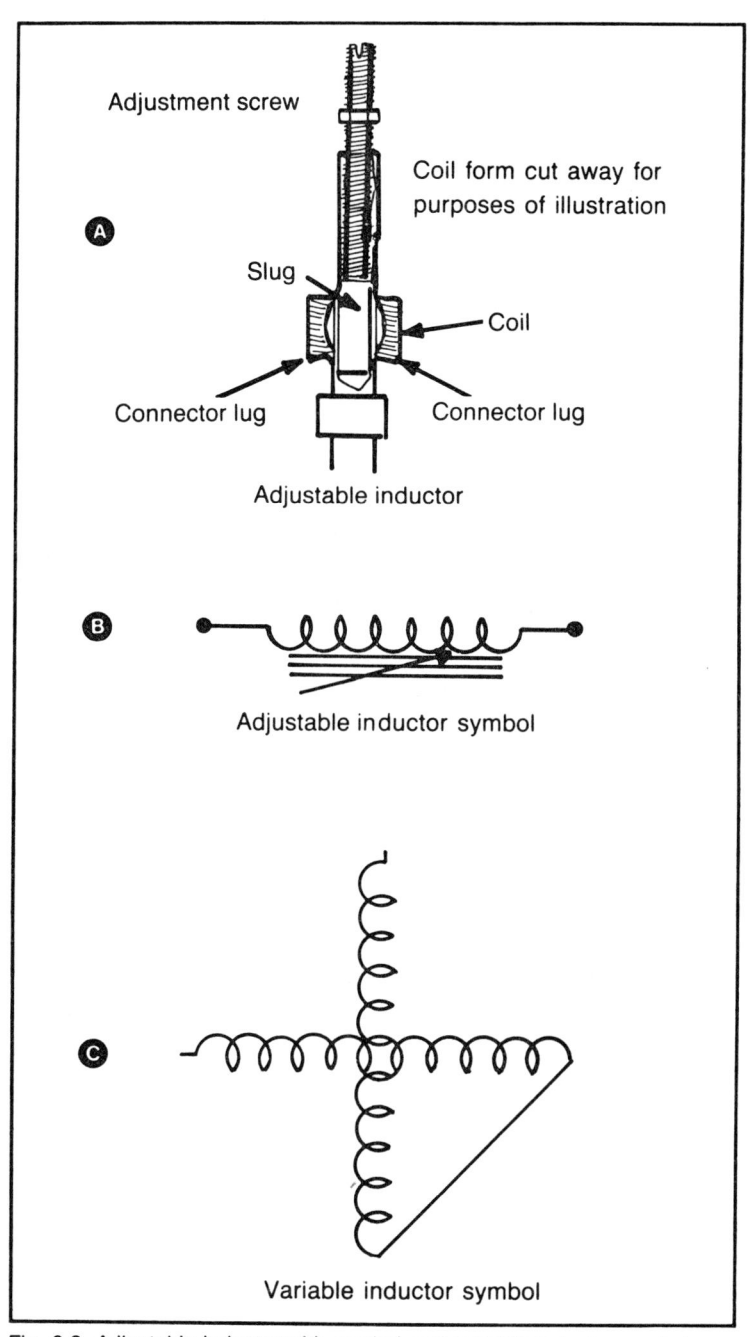

Fig. 9-3. Adjustable inductor with symbol and variable inductor symbol.

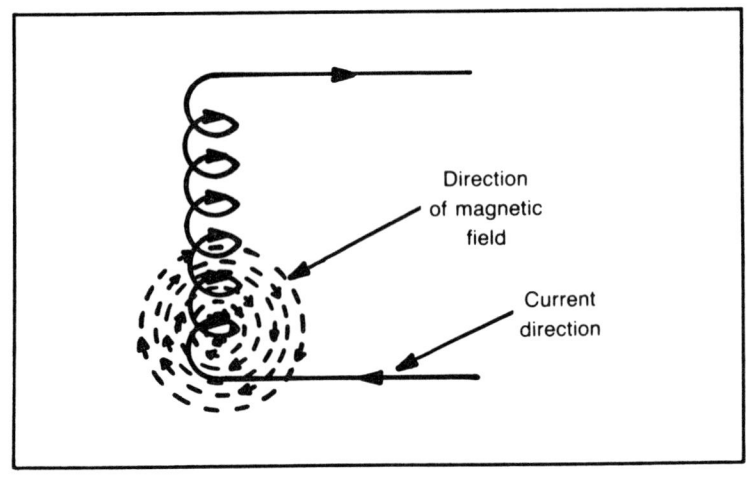

Fig. 9-4. Self-induction in a coil.

some or all of the neighboring turns on the coil, thereby inducing a voltage in these turns. The induced voltage causes a current flow that is in the opposite direction of the original current flow.

The reverse of this condition is also true. When current in a particular turn decreases, the field collapses and again cuts some or all of the neighboring turns, but in the reverse direction, trying to keep the original current from falling.

The basic unit of inductance is the *henry* (H). A circuit has an inductance of one henry when a current change of one ampere causes a counter emf of one volt to be developed. In practical work the henry is usually too large a unit; millihenry (mH) and microhenrys (μH) are generally used.

FACTORS AFFECTING INDUCTANCE

The inductance of an inductor is based only on its physical characteristics, such as the

- Number of turns of wire on the coil
- Spacing between turns
- Length and diameter of the coil
- Core material

Inductor design is itself an art, and many formulas are available for determining inductance. A basic formula giving the relationship of the physical factors affecting inductance is

$$L = \frac{1.26u\ AN^2}{1 \times 10^8}$$

where: L = inductance in henrys
1.26 = a constant
u = permeability of the core material
A = cross-sectional area of the core
N = number of turns of wire on coil
l = length of the coil

Inductive Effects in dc Circuits

The effects of inductance are apparent only when there is a change in the current flowing through a circuit. Placing an inductor in a direct current circuit, however, does effect the current. Before analyzing what happens to direct current in a practical in-

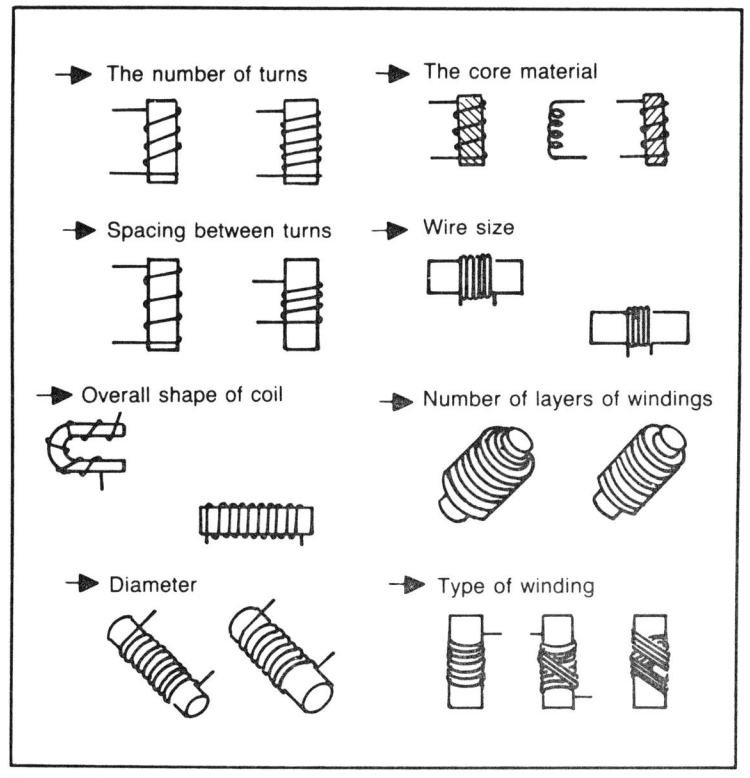

Fig. 9-5. Factors affecting inductance.

163

ductive circuit, we will look at the rise and fall of current in a pure resistive circuit.

Referring to Fig. 9-6, we see a dc pure-resistive circuit consisting of an emf, a switch, and a series resistor. When the switch is closed, current will flow from the negative terminal of the battery, through the resistor and switch, and then to the positive terminal of the battery. The very instant the switch is closed, the

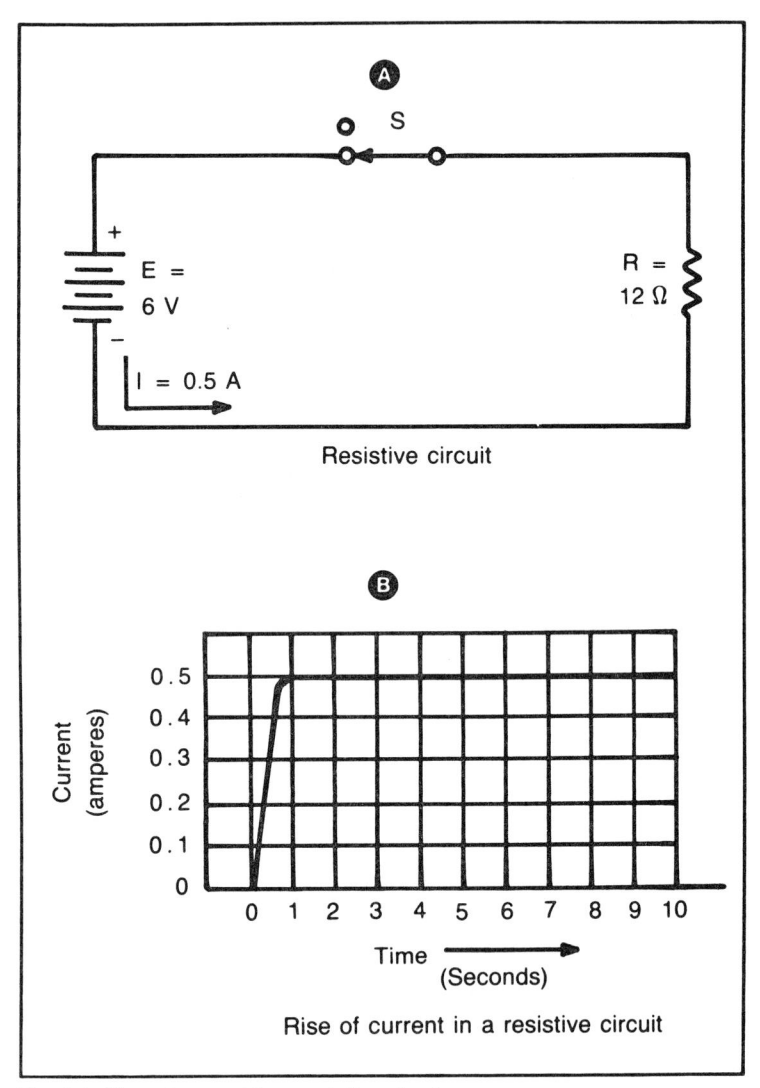

Fig. 9-6. Rise of current in a resistive circuit.

current in the circuit rises to its maximum value as shown by the current curve shown in Fig. 9-6B. In reality, the current does not change from zero to its maximum value instantaneously; however, the time interval is so short that it is considered to be instantaneous. The circuit current will now remain at its maximum level until the switch is opened, at which time the current will fall instantaneously back to zero.

In an electrical circuit containing both resistance and inductance, the rise and fall of current is quite different. Figure 9-7A shows a circuit consisting of a voltage source in series with an inductor, a resistor, and a switch. The resistor R includes all the resistance in the circuit; that is, the dc resistance of the inductor and the internal resistance of the battery are lumped into this one resistor. Usually the dc resistance of an inductor is not very large because the resistance depends only on the resistivity and length of wire used in the coil. When the switch is closed, the current *does not* instantaneously rise to its final steady value as in the resistive circuit because of the counter emf in the coil.

INDUCTIVE TIME CONSTANT

The inductive time constant is the ratio of the circuit inductance, L, to the circuit resistance, R. It represents the time in seconds required for the current to rise to 63.2 percent of its maximum value. The formula is

$$t = \frac{L}{R}$$

where: t = time in seconds for current to reach 63.2 percent of its maximum value
L = circuit inductance in henrys
R = circuit resistance in ohms

Using this formula, we can find the inductive time constant for the circuit in Fig. 9-7:

$$t = \frac{L}{R}$$
$$= \frac{2}{10}$$
$$= 0.2 \text{ sec}$$

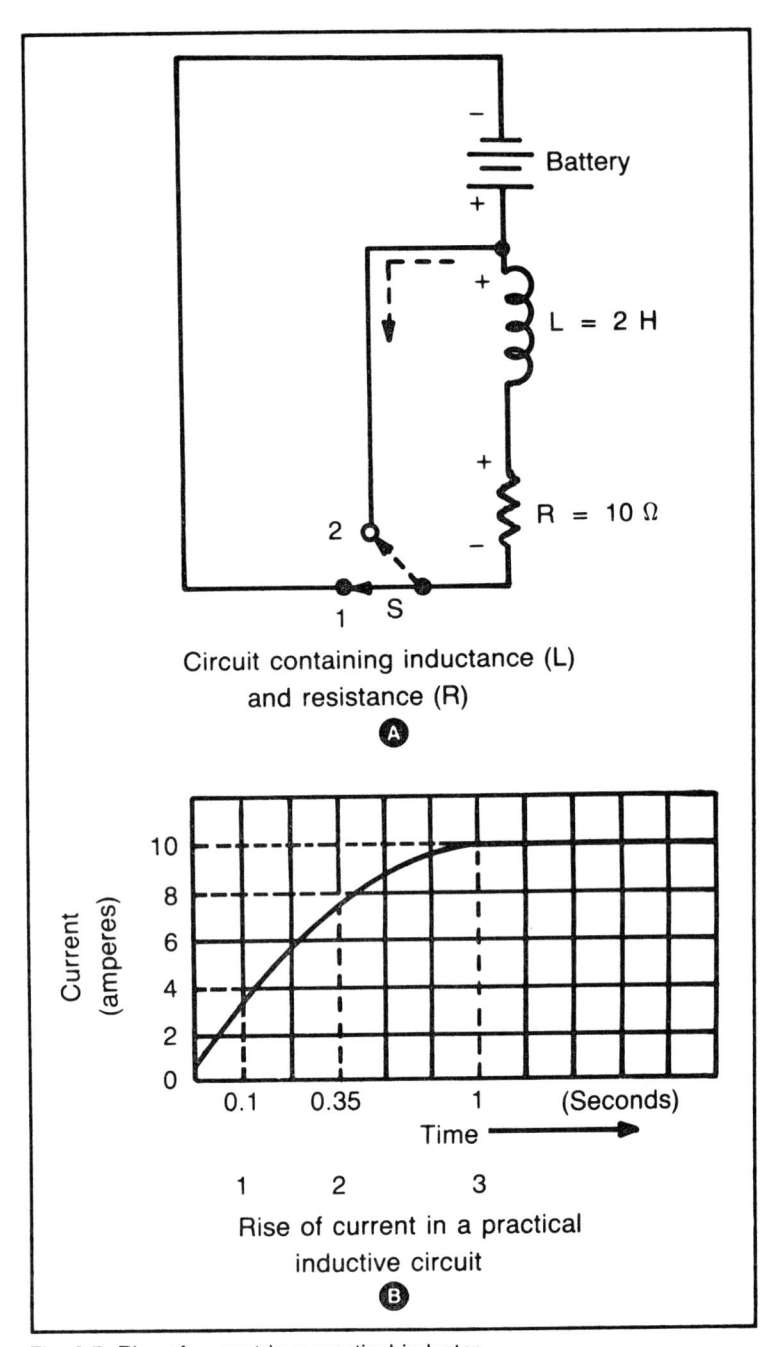

Fig. 9-7. Rise of current in a practical inductor.

Table 9-1 shows the effects of circuit inductance and resistance on the time constant.

MUTUAL INDUCTANCE

Because a changing current flowing through one turn of an inductor induces a voltage in adjacent turns of the same inductor, it is also possible for a voltage to be induced in a second inductor placed near the original inductor. When two inductors are so placed, they demonstrate an action called *mutual inductance*. When mutual inductance exists between two inductors, a change in current induces a voltage in the other.

Figure 9-8 shows this action. It is important to remember that mutual inductance takes place only while the current is changing and the magnetic lines of force surrounding L_1 are either expanding or collapsing.

The effects of mutual inductance are sometimes desirable; at other times they should be minimized. Therefore, we need to understand the factors affecting mutual inductance.

1. The number of turns of wire in both inductors
2. The relative position of the two inductors to each other
3. The permeability of the medium between the two inductors

You are already familiar with the first factor because the number of turns affects the amount of induced voltage. The second factor is illustrated in Fig. 9-9.

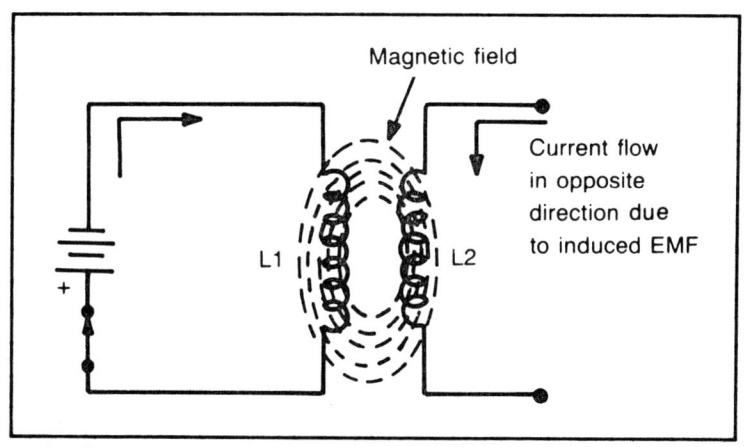

Fig. 9-8. Mutual inductance between two inductors.

167

Table 9-1. Wire-Size Chart.

Gauge (AWG or B & S)	Enamel	SSC	DSC SCC	DCC	Feet per Pound	Feet per Ohm
10	9.6		9.3	8.9	31.82	1001
11	10.7		10.3	9.8	40.12	794
12	12.0		11.5	10.9	50.59	629.6
13	13.5		12.8	12.0	63.80	499.3
14	15.0		14.2	13.8	80.44	396.0
15	16.8		15.8	14.7	101.4	314.0
16	18.9	18.9	17.9	16.4	127.9	249.0
17	21.2	21.2	19.9	18.1	161.3	197.5
18	23.6	23.6	22.0	19.8	203.4	156.5
19	26.4	26.4	24.4	21.8	256.5	124.2
20	29.4	29.4	27.0	23.8	323.4	98.5
21	33.1	32.7	29.8	26.0	407.8	78.11
22	37.0	36.5	34.1	30.0	514.2	61.95
23	41.3	40.6	37.6	31.6	648.4	49.13
24	46.3	45.3	41.5	35.6	817.7	38.96
25	51.7	50.4	45.6	38.6	1031	30.90
26	58.0	55.6	50.2	41.8	1300	24.50
27	64.9	61.5	55.0	45.0	1639	19.43
28	72.7	68.6	60.2	48.5	2067	15.41
29	81.6	74.8	65.4	51.8	2607	12.22
30	90.5	83.3	71.5	55.5	3287	9.691
31	101.0	92.0	77.5	59.2	4145	7.685
32	113	101	83.6	62.6	5227	6.095
33	127	110	90.3	66.3	6591	4.833
34	143	120	97.0	70.0	8310	3.833
35	158	132	104	73.5	10480	3.040
36	175	143	111	77.0	13210	2.411
37	198	154	118	80.3	16660	1.912
38	224	166	126	83.6	21010	1.516
39	248	181	133	86.6	26500	1.202
40	282	194	140	89.7	33410	0.954

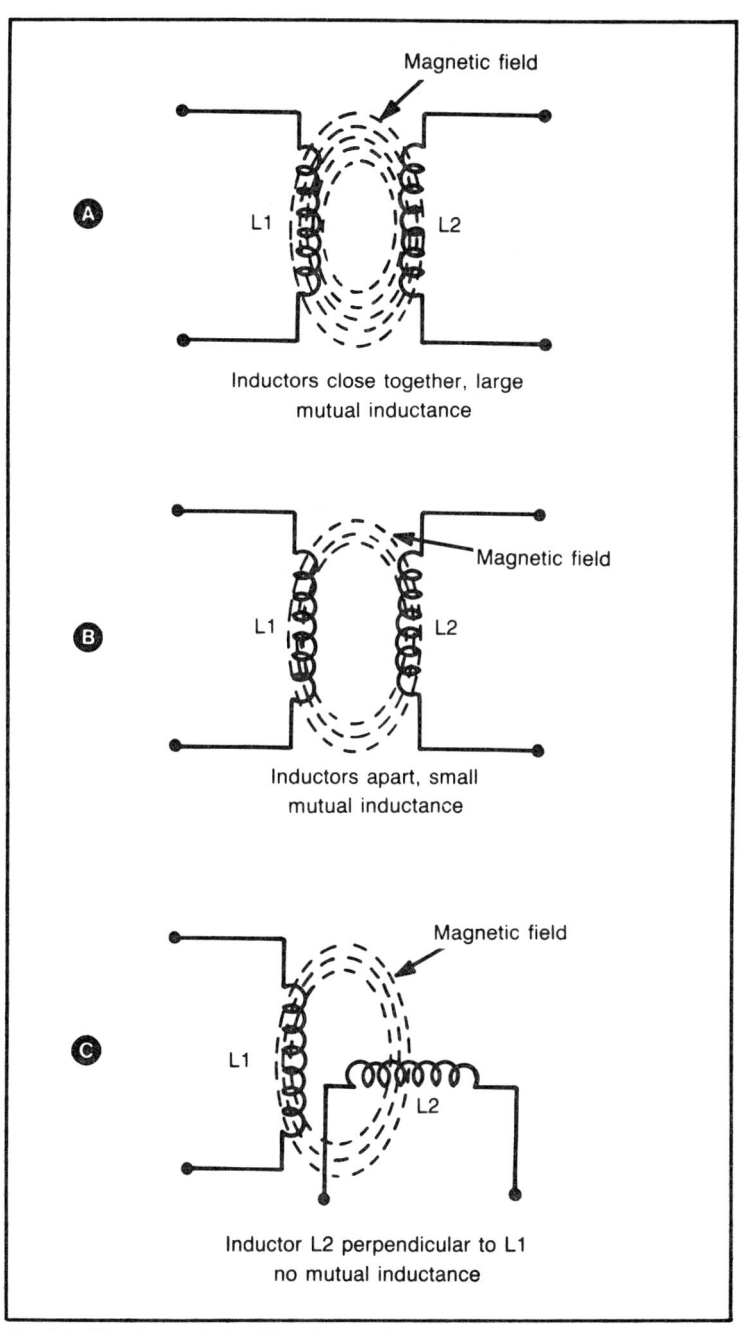

Fig. 9-9. Physical relationship of two inductors vs. mutual inductance.

CALCULATING TOTAL INDUCTANCE
Series-Connected Inductors

When inductors are connected in series, each one helps to oppose any change of current in the circuit. Because they are connected in series, the same current flows through each inductor. Therefore, the sum of the voltage drops across the individual series-connected inductors will be equal to the total circuit voltage.

Figure 9-10 shows a circuit containing inductors L_1, L_2, and L_3 connected in series. To determine the equivalent total inductance of the circuit, simply add the individual inductances.

$$L_t = L_1 + L_2 + L_3$$

Parallel-Connected Inductors

Calculating total inductance for parallel-connected inductors is similar to calculating total resistance for parallel-connected resistors. When resistors are connected in parallel, the smallest resistor shorts out the larger resistors. Most of the circuit current flows through the smallest resistor because it offers the least opposition to current flow. When inductors are connected in parallel, as shown in Fig. 9-11, and the current is either increased or decreased, each inductor individually opposes this change in current. Most of the circuit current then chooses the path that offers the least opposition to this flow. The formula is

$$L_t = \frac{1}{\frac{1}{L_1} + \frac{1}{L_2} + \frac{1}{L_3} + \ldots}$$

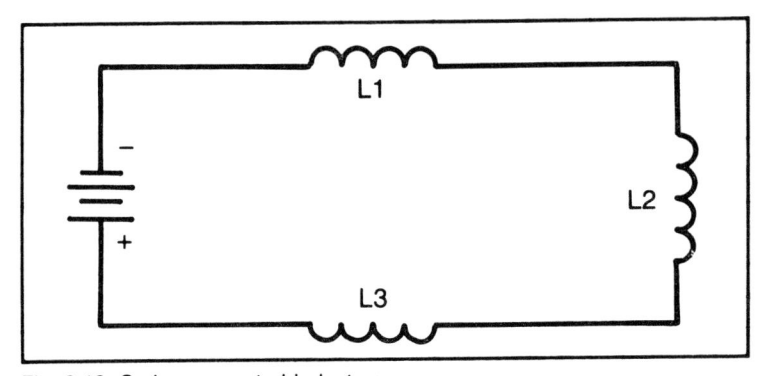

Fig. 9-10. Series-connected inductors.

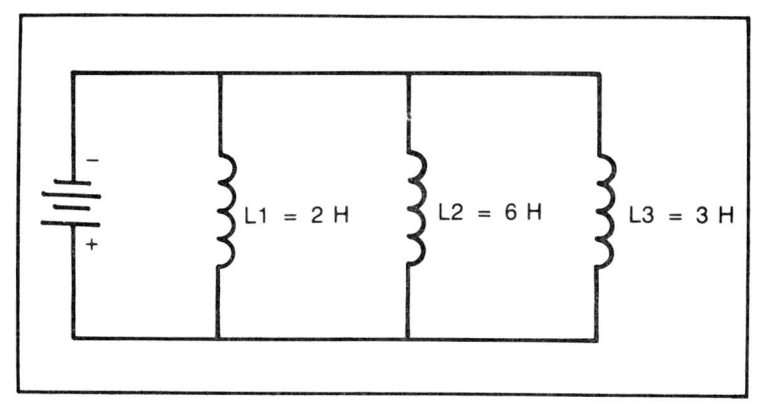

Fig. 9-11. Parallel-connected inductors.

Substituting the values in Fig. 9-11 into the equation and solving, we find that the circuit has a total inductance of 1 H. This is less than the smallest inductance value in the circuit.

For only two parallel-connected inductors, we can use the product-over-sum method:

$$L_t = \frac{L_1 \times L_2}{L_1 + L_2}$$

INDUCTOR CONSTRUCTION

In electrical and electronic equipment, inductors are used for many purposes and come in many sizes and shapes, depending on their function.

Up to this point, inductors have been referred to as simple, one-layer, air core or iron core types. The equation

$$L = \frac{1.26u \ AN^2}{1 \times 10^8}$$

was presented to illustrate how permeability, cross-sectional area, length of the core, and number of turns affected the inductance of coils. In actual applications, however, the dimensions of inductors seldom conform to the conditions required by this simple equation.

Inductor construction is classified into four general groups.

1. Single-layer windings

2. Multilayer windings
3. Flat-spiral or pancake windings
4. Toroidal windings

A separate equation is used for each group.

Single-Layer Windings

Figure 9-12 shows a typical single-layer winding inductor. This type of coil is generally used in solenoids and certain types of chokes. To calculate the inductance for this type use the equation

$$L = \frac{(rN)^2}{9r + 10l}$$

where: L = coil inductance in microhenrys
r = mean coil radius in inches
N = total number of turns on the coil
l = coil length in inches

For example, let us calculate the inductance of a single-layer coil with 150 turns of wire, a length of 1 in and mean radius of 0.9 in.

$$L = \frac{(rN)^2}{9r + 10l}$$

$$= \frac{(0.9 \times 150)^2}{(9 \times 0.9) + (10 \times 1)}$$

$$= 1006.9 \ \mu H$$

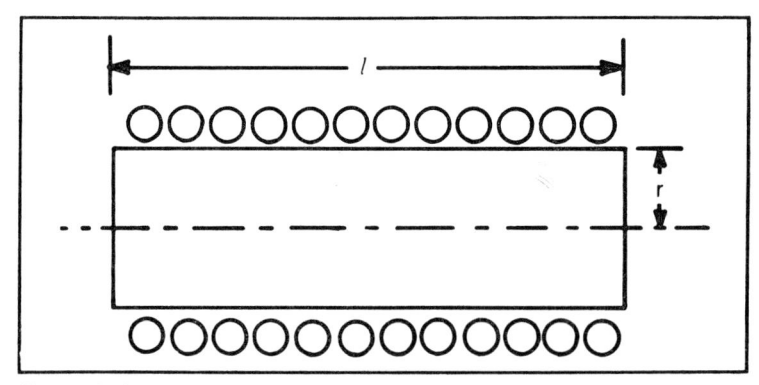

Fig. 9-12. Axial section view of a single-layer winding.

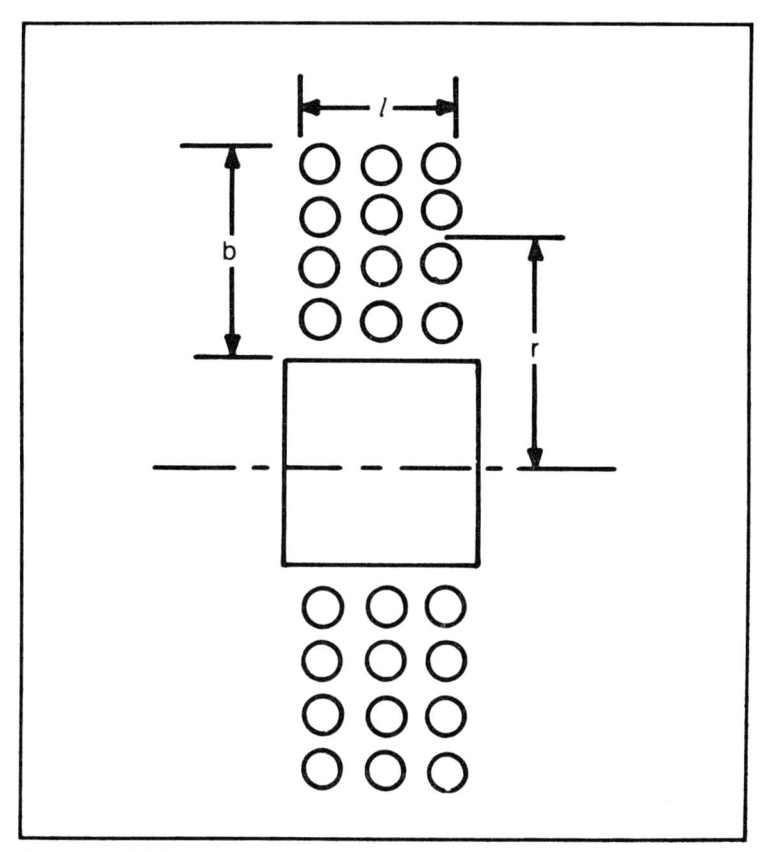

Fig. 9-13. Multilayer coil winding.

Multilayer Windings

Figure 9-13 shows a view of a multilayer winding. This type of construction is widely used in choke coils and transformers. The equation for calculating the inductance of this type of coil is

$$L = \frac{0.8(rN)^2}{6r + 9l + 10b}$$

where: L, r, N, and l are as in a single-layer winding
b = depth of coil in inches

Flat-Spiral or Pancake Windings

Figure 9-14 shows a view of a pancake coil winding. It is often wound with rectangular wire or thin metal ribbon rather than cir-

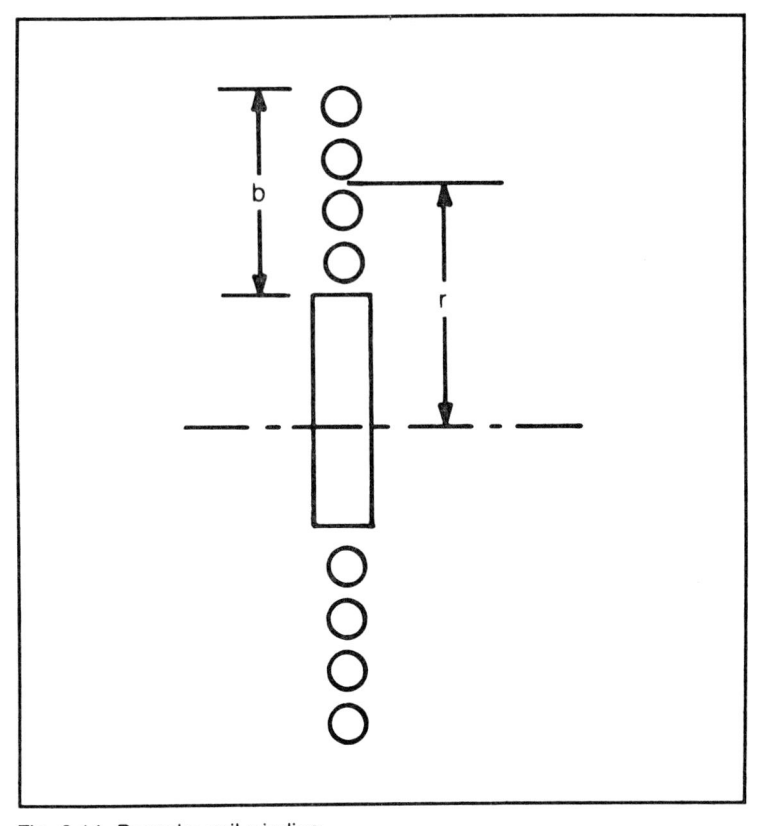

Fig. 9-14. Pancake coil winding.

cular wire. The equation for determining the inductance of a pancake winding is

$$L = \frac{(rN)^2}{8r + 11b}$$

Toroidal Coil Windings

Figure 9-15 illustrates the construction of a toroidal coil winding. This type has a completely closed circular or rectangular core. It is called "toroidal" because the turns are wound on a torus or a ring. Toroidal coils are used in computer memory systems and filter chokes. Because they are wound on either single-layer or multilayer forms, the equations for calculations are the same as already discussed.

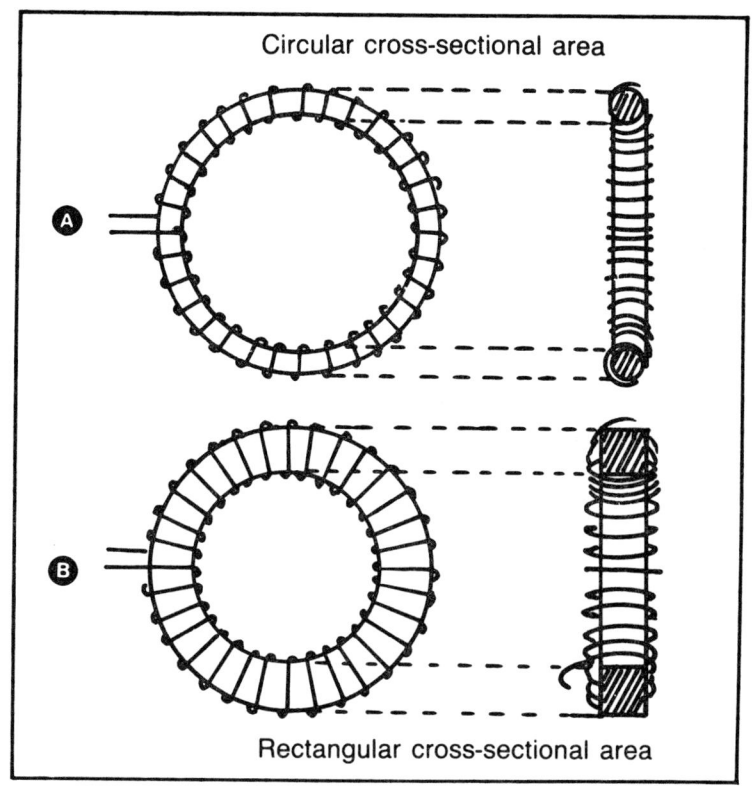

Fig. 9-15. Toroidal coil winding.

Wire size must be considered when designing a coil. Table 9-2 lists various wire sizes and types and shows how many turns per inch may be obtained.

Noninductive Windings

Sometimes the effects of inductance are not wanted—for ex-

Table 9-2. Effects of Circuit Inductance and dc Resistance on Time Constant.

Inductance L	Resistance R	TimeConstant $t = \frac{L}{R}$
↑	→	↑
↓	→	↓
→	↑	0↓
→	0↓	↑

175

ample, in wirewound resistors for precision circuits. To obtain little or no inductance, we would have to wind the resistance coil in such a manner that the inductive effect created by one turn cancels, and in turn is cancelled by, the inductive effect of the adjacent turn. When wound in this manner, the inductor is said to be *noninductively wound*.

One method is to first double the wire to be used. Figure 9-16 shows this. In inductors of this type the current flows in one direction through one-half of the coil and in the opposite direction through the other half. The flow of current in opposite directions within the same coil causes the magnetic effects of each turn to cancel each other.

USES FOR INDUCTORS

Inductors have many uses: choke or filter coils, transformers, tuned circuits, relays and solenoids, and special applications. Many are strictly for ac applications, but some important ones are for dc uses. An important example is the automotive ignition system, which is still in use in many older cars (see Fig. 9-17).

The main shaft of the engine controls a cam that activates a set of breaker points that open and close the primary circuit. The breaker points are included in the distributor housing and function

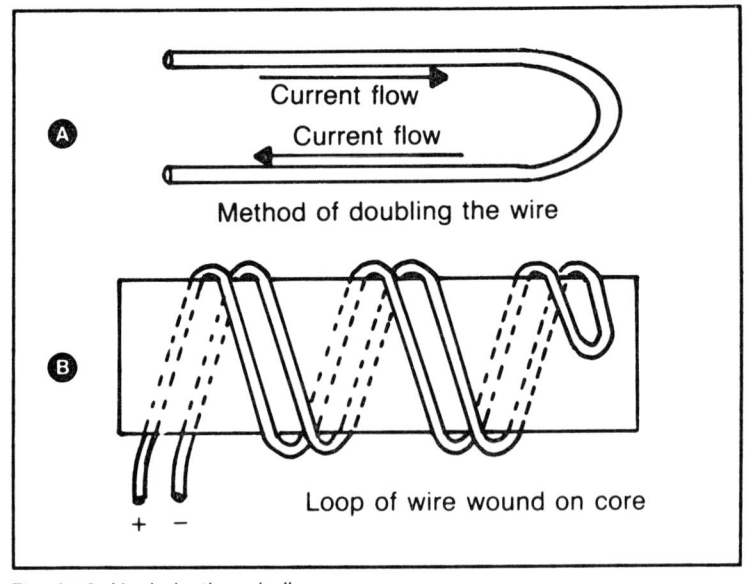

Fig. 9-16. Noninductive winding.

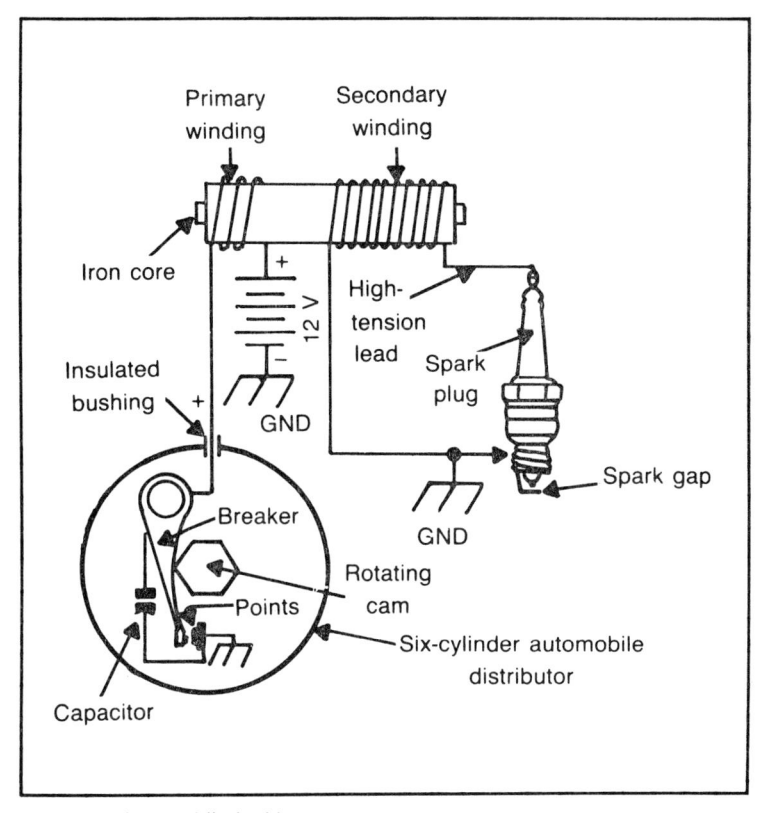

Fig. 9-17. Automobile ignition system.

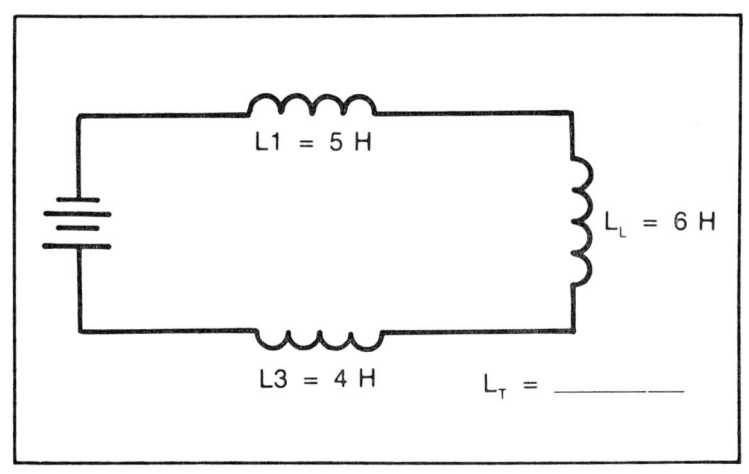

Fig. 9-18. Series inductive circuit.

177

as a properly timed and mechanically operated switch controlling the current flowing through the primary winding of the coil. Notice that the primary winding has only a few turns of heavy wire, whereas the secondary winding has many turns of fine wire on the same iron core. This coil is designed to have a high mutual inductance, which in turn causes a high voltage to be induced in the secondary.

During the time that the points are closed, the 12-V battery sends a current through the primary winding, creating a magnetic field that links the secondary winding. When the points are opened by the action of the cam, the current flow in the primary is suddenly interrupted, causing a rapid collapse of the magnetic field. This rapid collapse causes a very high voltage (10-50 kV) to be induced in the secondary. It must be high enough to jump the gap in the spark plug to ignite the gasoline-air mixture in the engine.

The capacitor placed across the points aids the collapse of the magnetic field and prevents sparking across the points.

QUIZ

1. Inductance is the property that opposes any change in

    A. Voltage
    B. Current
    C. Resistance

2. Name the four basic physical factors that affect inductance.
3. Draw the schematic symbol for an air core fixed inductor.
4. The basic unit of inductance is the _____.

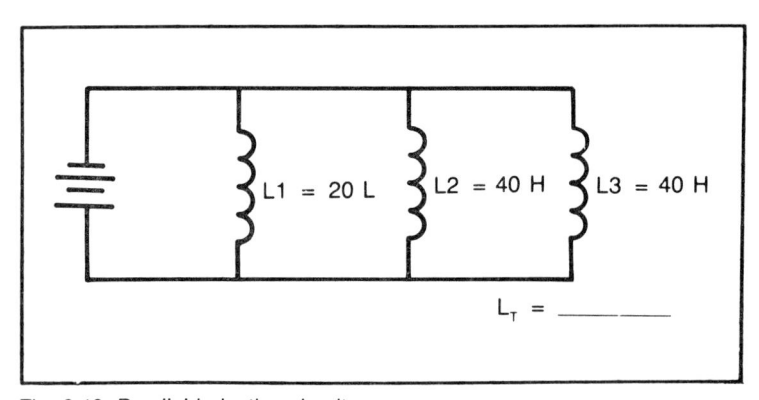

Fig. 9-19. Parallel inductive circuit.

5. A circuit containing 5 Ω of resistance and 5 H of inductance has an inductive time constant of _____ sec.

6. How long does it take current to reach its full value in an inductive circuit?

7. The value of L_t in Fig. 9-17 is _____.

8. The vlaue of L_t in Fig. 9-18 is _____.

9. AWG 19 enamel wire is capable of _____ turns per linear inch.

10. A wirewound resistor is generally made up with _____ inductive windings.

Appendix A

Answers to Quizzes

Chapter 1
1. The free electron
2. Nucleus and electrons
3. The rate of doing work
4. A form of static electricity
5. A material containing many free electrons
6. 746 W
7. The smallest particle to which an element can be reduced and still retain its original characteristics
8. A material containing few or no free electrons
9. Like charges repel, unlike charges attract
10. The ability or capacity to do work
11. Percent Eff $= \dfrac{\text{Power Out}}{\text{Power In}} \times 100$
12. Matter is anything that has mass and occupies space
13. Silver
14. Coal
15. Glass

Chapter 2
1. Negative, positive
2. Emf
3. Six
4. Chemical and magnetic

5. Ampere
6. Ohm
7. Carbon
8. Voltage (emf)
9. Current
10. Resistance

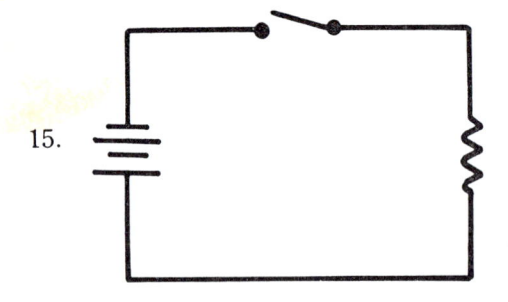

11. ⊣|ı|ı|ı|⊢

12. (fuse symbol)

13. (switch symbol)

14. (resistor symbol)

15. (circuit diagram with battery, switch, and resistor)

Chapter 3

1. $I = \dfrac{E}{R}$

   $E = I \times R$

   $R = \dfrac{E}{I}$

2. Current is also doubled

3. 285 Ω
4. 75 Ω
5. 96 V
6. 10 A
7. R_t = 16 Ω, R_3 = 8 Ω, R_2 = 3 Ω, E_1 = 15 V, E_2 = 9 V
8. R_t = 12 Ω, R_3 = 4 Ω, E_1 = 2 V, I_1 = 1 A
9. R_t = 4 Ω

Chapter 4
1. Like poles repel, unlike poles attract
2. True
3. North to south
4. South to north
5. No
6. Maxwell
7. Gilbert
8. Reluctance
9. Horseshoe magnet
10. Magnetic lines of force

Chapter 5
1. The primary cell is not rechargeable. The secondary cell is rechargeable.
2. Voltage
3. Current
4. Hydrogen
5. 10
6. Constant current, constant voltage
7. Buckling of the plates
8. Electrolyte
9. Specific gravity
10. Hydrometer

Chapter 6
1. Yes
2. 2 W
3. Rheostats and potentiometers
4. 95 to 105 Ω
5. 50,000 Ω ± 5 percent
6. Negative
7. Thermistor
8. Varistor
9. Wirewound

Chapter 7
1. Series
2. Parallel
3. Ohms per volt
4. Short circuit
5. Open circuit
6. Glancing at a meter pointer from either side
7. Digital voltmeter
8. No
9. Yes
10. Yes

Chapter 8
1. A
2. 63.2
3. 4500 pF
4. Plate area; distance between plates; type of dielectric
5. 60 µF
6. 13 µF
7. 0.000550 F
8. 5
9. mica
10. Farad

Chapter 9
1. B
2. Turns of wire, spacing between turns, shape of coil, material of the core
3. ⎯⎯⏝⏝⏝⎯⎯
4. Henry
5. 1 sec
6. 5 time constants
7. 13 H
8. 10 H
9. 26.4
10. Non

Appendix B

Dc Formulas

Ohm's Law

$E = IR$

$I = \dfrac{E}{R}$

$R = \dfrac{E}{I}$

Power

$P = IE$

$P = I^2 R$

$P = \dfrac{E^2}{R}$

$I = \dfrac{P}{E}$

$E = \dfrac{P}{I}$

Series Circuits

$R_t = R_1 + R_2 + R_3 + \ldots$
$E_t = E_1 + E_2 + E_3 + \ldots$
$P_t = P_1 + P_2 + P_3 + \ldots$
$L_t = L_1 + L_2 + L_3 + \ldots$

$$C_t = \dfrac{1}{\dfrac{1}{C_1} + \dfrac{1}{C_2} + \dfrac{1}{C_3} + \ldots}$$

or

$$C_t = \frac{C_1 \times C_2}{C_1 + C_2}$$

I_t = Current is the same in all parts of a series circuit

Parallel Circuits

$$R_t = \frac{1}{\frac{1}{R_1} + \frac{1}{R_2} + \frac{1}{R_3} + \ldots}$$

or

$$R_t = \frac{R_1 \times R_2}{R_1 + R_2}$$

$$L_t = \frac{1}{\frac{1}{L_1} + \frac{1}{L_2} + \frac{1}{L_3} + \ldots}$$

or

$$L_t = \frac{L_1 \times L_2}{L_1 + L_2}$$

$C_t = C_1 + C_2 + C_3 + \ldots$
$I_t = I_1 + I_2 + I_3 + \ldots$
E_t = Voltage is the same in each branch of a parallel circuit

Appendix C

Electrical Device Reference Designations

Device	*Abbreviation*
Alarm	DS
Amplifier	A
Amplifier, rotating	G
Annunciator	DS
Antenna	E
Arrestor, lightning	E
Assembly	A
Attenuator	AT
Audible signaling device	DS
Autotransformer	T
Battery	BT
Bell	DS
Blower, fan, motor	B
Board, terminal	TB
Breaker, circuit	CB
Buzzer	DS
Cable	W
Capacitor	C
Cell, aluminum or electrolyte	E
Cell, light sensitive	V
Choke	L

Device	Abbreviation
Circuit breaker	CB
Coil, hybrid	HY
Coil	L
Coil, repeating	T
Computer	A
Connector receptacle (wall or panel)	J
Connector at end of wire or cable	P
Contact, electrical	E
Contactor	K
Directional coupler	DS
Crystal detector	CR
Crystal diode	CR
Crystal piezoelectric	Y
Cutout fuse	F
Cutout thermal	S
Detector, crystal	CR
Indicating device	DS
Dipole antenna	E
Disconnecting device	S
Electron tube	V
Exciter	G
Fan	B
Filter	FL
Fuse	F
Generator	G
Handset	HS
Head, erasing	PU
Heater	HR
Horn, howler	LS
Indicator	DS
Inductor	L
Instrument	M
Insulator	E
Interlock	S
Junction hybrid	HY
Key switch	S
Lamp, pilot	DS
Lamp, signal	DS

Device	Abbreviation
Line, delay	DL
Loop antenna	E
Magnet	E
Meter	M
Microphone	MK
Mode transducer	MT
Modulator	A
Motor	B
Motor generator	MG
Nameplate	N
Oscillator	Y
Oscilloscope	M
Pad	AT
Part, misc	E
Path, guided transmission	W
Phototube	V
Pickup head	PU
Potentiometer	R
Power supply	A
Receiver telephone	HT
Receptacle (jack)	J
Rectifier	CR
Regulator, voltage	VR
Relay	K
Repeater (telephone)	RP
Resistor	R
Rheostat	R
Selenium cell	CR
Shunt	R
Solenoid	L
Speaker	LS
Speed regulator	S
Strip, terminal	TP
Subassembly	A
Switch	S
Terminal board	TB
Test point	TP
Thermistor	RT
Thermocouple	TC
Thermostat	S

Device	Abbreviation
Timer	M
Transducer	MT
Transistor	Q
Transmission path	W
Tube, electron	V
Varistor, asymmetrical	CR
Varistor, symmetrical	RV
Voltage regulator	VR
Waveguide	W
Winding	L
Wire	W

Appendix D

Abbreviations for Electrical Devices

Device	Abbreviation
Adaptor	ADPT
Air circuit breaker	ACB
Alternating current	AC
Alternating current, volts	VAC
Aluminum	AL
Ammeter	AM
Ampere	AMP
Amplifier	AMPL
Antenna	ANT
Armature	ARM
Arrestor	ARR
Attenuator, attenuation	ATTEN
Audio frequency	AF
Automatic gain control	AGC
Battery	BAT
Beat frequency oscillator	BFO
Bottom	BOT
Cabinet	CAB
Capacitor	CAP
Cathode ray tube	CRT
Circuit	CKT
Coaxial	COAX

Device	Abbreviation
Collector	COLL
Compress	COMP
Condenser	COND
Conductor	COND
Conduit	CND
Counterclockwise	CCW
Cycles per second	CPS [Note: replaced by Hz (Hertz)]
Decibel	DB
Diameter	DIA
Direct current	DC
Double-pole, double-throw	DPDT
Double-pole, single-throw	DPST
Drawing	DWG
Dynamometer	DYNO
Dynamotor	DYNM
Electric horsepower	EHP
Electrolytic	ELECT
Electronic Industries Assn.	EIA
Engineer	ENGR
Engineering	ENGRG
Escutcheon	ESC
Exciter	EXC
Field reversing	FFR
Flat head	FH
Fluorescent	FLUOR
Fuse	FZ
Gauge	GA
Germanium	Ge
Grommet	GROM
Guided missile	GM
Heater	HTR
Heat treat	HT TR
Hertz	HZ
High frequency	HF
High-frequency oscillator	HFO
High voltage	HV
Horizon, horizontal	HORIZ
Ignition	IGN
Indicator	IND
Induction-capacitance	LC
Induction	IND

Device	Abbreviation
Instrument	INST
Intermediate frequency	IF
Junction box	JB
Knockout	KO
Kilohertz	KHZ
Kilohm	K
Kilovolt	KV
Kilovolt ampere	KVA
Kilowatt	KW
Kilowatt hour	KWH
Lighting	LTG
Low frequency	LF
Low voltage	LV
Magnetic amplifier	MAG AMPL
Magnetic modulator	MAG MOD
Manual	MAN
Master switch	MS
Medium frequency	MF
Mega (10^6)	MEG
Megahertz	MHZ
Megohm	MEGO
Meter	M
Metering	MTRG
Missile	MSL
Modify	MOD
Modulator	MOD
Motor	MOT
Mounting	MTG
Multiplex	MUX
Multivibrator	MVB
National electrical code	NEC
Not to scale	NTS
Oil circuit breaker	OCB
Oscillator	OSC
Overload	OVLD
Phase	PH
Potentiometer	POT
Power supply	PWR SUP
Radar	RDR
Radio	RAD
Receptacle	RECP

Device	Abbreviation
Reference	REF
Resistance	RES
Resistance-capacitance	RC
Resistor	RES
Roundhead	RH
Schedule	SCH
Screw	SCR
Secondary	SEC
Selector	SEL
Selenium	Se
Servomechanism	SERVO
Signal	SIG
Single-pole double-throw	SPDT
Single-pole single-throw	SPST
Solenoid	SOL
Speaker	SPKR
Specification	SPEC
Suppressor	SUPPR
Switch	SW
Switchboard	SWBD
Synchronous	SYN
Tachometer	TACH
Technical manual	TM
Telemeter	TLM
Terminal	TERM
Test switch	TSW
Thermistor	TMTR
Thermocouple	TC
Three conductor	3/C
Three phase	3 PH
Time delay	TD
Transceiver	XCVR
Transformer	XFMR
Transistor	TSTR
Transmitter	XMTR
Tuning	TUN
Twisted	TW
Ultrahigh frequency	UHF
Unfused	UNF
Vacuum tube	VT
Vacuum tube voltmeter	VTVM

Device	*Abbreviation*
Var hour meter	VRH
Variable-frequency oscillator	VFO
Very high frequency	VHF
Very low frequency	VLF
Video	VID
Volt	V
Voltage regulator	VR
Voltmeter	VM
Volume	VOL
Watt	W
Watt hour	WHR
Watt-hour meter	VHM
Wattmeter	WM
Wirewound	WW

Appendix E

Wire Gauge Sizes and Current Capacity

AWG no.	Nominal Diameter (in) bare	enameled	Ampere Rating
10	0.1019	0.1039	30
12	0.0808	0.0827	20
14	0.0641	0.0661	15
16	0.0508	0.0526	10
18	0.0403	0.0419	7
20	0.0320	0.0335	5
22	0.0253	0.0267	3
24	0.0201	0.0213	
26	0.0159	0.0170	
28	0.0126	0.0135	
30	0.0100	0.0106	
32	0.0080	0.0087	
34	0.0063	0.0069	
36	0.0050	0.0055	
38	0.0040	0.0044	
40	0.0031	0.0035	

Appendix F

Symbols for Electrical-Electronic Devices

Adjustable

continuously adjustable, variable

Amplifier

general

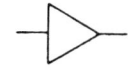

with two inputs

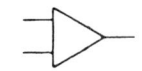

with two outputs

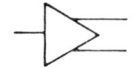

with adjustable gain

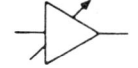

with associated attenuator

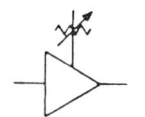

with associated power supply

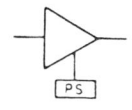

with external feedback path

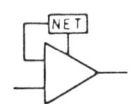

Amplifier Letter Combinations

(May be used with amplifier symbols if needed for explanation).

BDG	bridging
BST	booster
CMP	compression
DC	direct current
EXP	expansion
LIM	limiting
MON	monitoring
PGM	program
PRE	preliminary
PWR	power
TRQ	torque

Antenna

general

dipole

loop

loop antenna (alternate symbol)

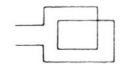

199

counterpoise, antenna

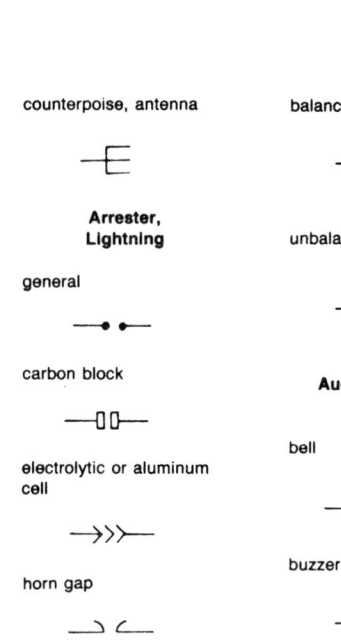

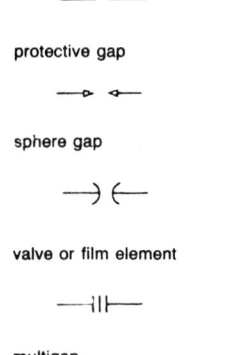

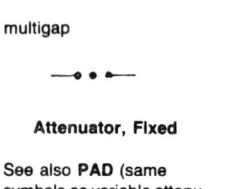

Arrester, Lightning

general

carbon block

electrolytic or aluminum cell

horn gap

protective gap

sphere gap

valve or film element

multigap

Attenuator, Fixed

See also **PAD** (same symbols as variable attenuator without adjustment arrow.)

Attenuator Variable

general

balanced

unbalanced

Audible Signaling Device

bell

buzzer

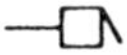

loudspeaker

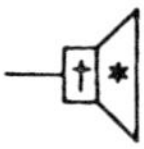

Loudspeaker Letter Combinations

HN	horn, electrical
HW	howler
LS	loudspeaker
SN	siren
EM	electromagnetic with moving coil
EMN	electromagnetic, moving coil and neutralized winding
MG	magnetic armature
PM	permanent magnet

sound, telegraph

Battery

one cell

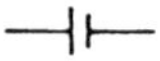

multicell

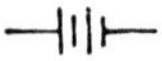

multicell with taps

multicell with adjustable lap

Capacitor

general

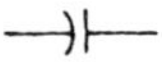

polarized

adjustable or variable

adjustable or variable with mechanical linkage

continuously adjustable or variable differential

phase shifter

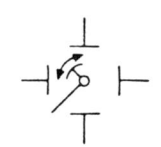

split stator

feed-through

Cell Photosensitive

asymmetrical photoconductive transducer

symmetrical photoconductive transducer

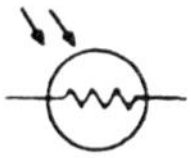

photovoltaic transducer

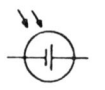

Circuit Breaker

general

Circuit Element

general

Letter Combinations for Circuit Elements

CB	circuit breaker
DIAL	telephone dial
EQ	equalizer
FAX	facsimile set
FL	filter
FL-BE	filter, band elimination
FL-BP	filter, band pass
FL-HP	filter, high pass
FL-LP	filter, low-pass
NET	network
PS	power supply
RU	reproducing unit
RG	recording unit
TEL	telephone station
TPR	telephone
TTY	teletypewriter

Additional Letter Combinations

(Specific graphical symbols preferred.)

AR	amplifier
AT	attenuator
C	capacitor
HS	handset
I	indicating lamp
L	inductor
LS	loudspeaker
J	jack
MIC	microphone
OSC	oscillator
PAD	pad
P	plug
HT	receiver, headset
K	relay
R	resistor
S	switch
T	transformer
WR	wall receptacle

Ground

earth ground

chassis connection

common connections

(Identifying marks to denote points tied together shall replace (*) asterisks.)

Clutch; Brake

clutch disengaged when operating means deenergized

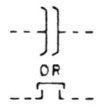

clutch engaged when operating means deenergized

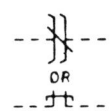

brake applied when operating means energized

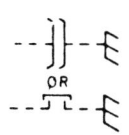

201

brake released when
operating means energized

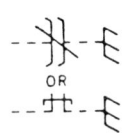

coaxial connector with
outside conductor carried
through

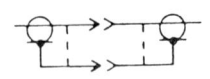

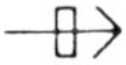

choke (rectangular
waveguide)

two-conductor switchboard
jack

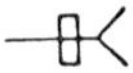

**Contact,
Electrical**

fixed contact for jack, key
or relay

→ OR ↑ OR →

**Coil Operating
(relay)**

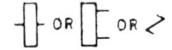

two-conductor switchboard
plug

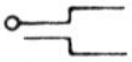

fixed contact for switch

○ OR →

female contact
(convenience outlets and
mating connectors)

fixed contact for
momentary switch

dot shows inner end of
winding

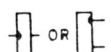

sleeve

male contact (convenience
outlets and mating
connectors)

**Connection,
Mechanical
(Interlock)**

moving contact,
adjustable

→ OR ↑

two-conductor non-
polarized connector
with female contacts

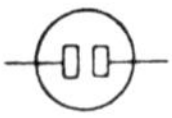

with fulcrum

moving contact, locking

Connector

female contact

two-conductor polarized
connector with male
contacts

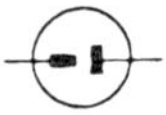

moving contact, nonlocking

male contact

segment, bridging contact

⌐ OR ⌐

Waveguide Flanges

separable connectors
(engaged)

vibrator reed

mated (general)

separable connectors
(alternate symbol)

vibrator split reed

plain (rectangular
waveguide)

rotating contact

202

closed contact, break

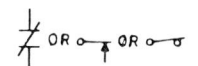

open contact, make

transfer

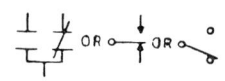

make-before-break

open contact with time closing or time-delay closing

closed contact with time opening or time-delay opening

time-sequential-closing

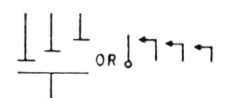

Core

air core

No Symbol

magnetic core of inductor or transformer

core of magnet

Counter, Electromechanical

Coupler, Directional

general

E-plane aperture coupling, 30-dB loss

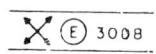

loop coupling, 30-dB loss

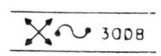

probe coupling, 30-dB loss

resistance coupling, 30-dB loss

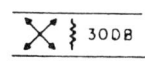

Coupling

(by aperture of less than waveguide size)

(Replace asterisk (*) by E, H or HE depending upon type of coupling to guided transmission path.)

Delay-Function

general

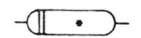

tapped delay

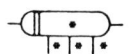

(Replace asterisk (*) with value of delay.)

Direction of Flow

one way

both ways

Discontinuity

equivalent series element

capacitive reactance

inductive reactance

inductance-capacitance circuit, infinite reactance at resonance

203

inductance-capacitance circuit, zero reactance at resonance

resistance

equivalent shunt element

capacitive susceptance

conductance

inductive susceptance

inductance-capacitance circuit with infinite susceptance at resonance

inductance-capacitance circuit with zero susceptance at resonance

Electron Tube

directly heated cathode, heater

indirectly heated cathode

cold cathode (including ionically heated cathode)

photocathode

pool cathode

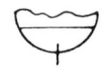

ionically heated cathode with supplementary heating

grid

deflecting electrode

ignitor

excitor

anode or plate

target or x-ray anode

dynode

composite anode-photocathode

composite anode-cold cathode

composite anode-ionically heated cathode with supplementary heating

shield, within envelope and connected to a terminal

outside envelope of x-ray tube

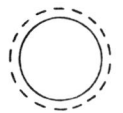

coupling by loop

resonator, cavity type-single-cavity envelope with grid electrodes

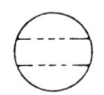

resonator-double cavity envelope with grid electrodes

204

multicavity magnetron anode and envelope

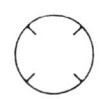

envelope

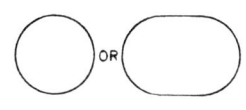

split envelope

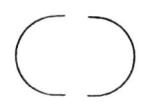

gas-filled envelope

basing orientation, tubes with keyed bases

basing, tubes with bayonets, bosses, or other reference points

base terminals

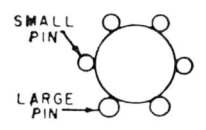

envelope terminals

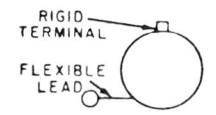

triode with directly heated cathode and envelope connection to base terminal

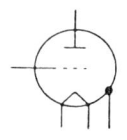

pentode

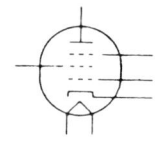

twin triode equipotential cathode

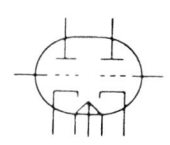

cold-cathode voltage regulator

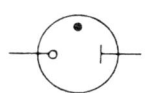

vacuum phototube

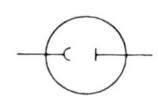

multiplier phototube

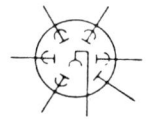

cathode-ray tube, electrostatic deflection

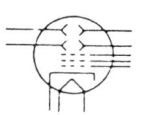

cathode-ray tube, magnetic deflection

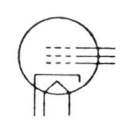

mercury-pool tube with ignitor and control grid

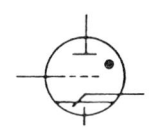

mercury-pool tube with excitor, control grid and holding anode

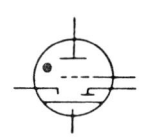

single-anode pool-type vapor rectifier with ignitor

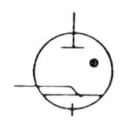

six-anode metal-tank pool-type rectifier with excitor

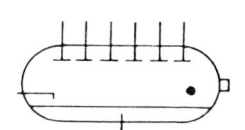

205

resonant magnetron with coaxial output

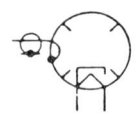

resonant magnetron with permanent magnet

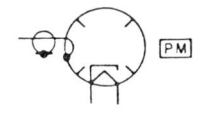

transit-time magnetron

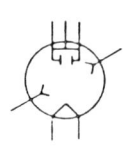

tunable magnetron

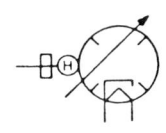

reflex klystron, integral cavity

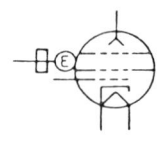

double-cavity klystron, integral cavity

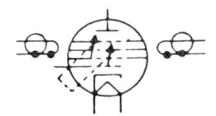

transmit-receive (t-r) tube

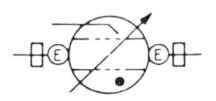

x-ray tube with directly heated cathode and focusing grid

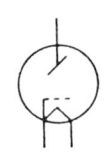

x-ray tube with control grid

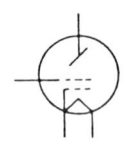

x-ray tube with grounded shield

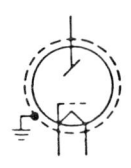

double-focus x-ray tube with rotating anode

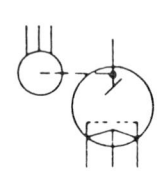

x-ray tube with multiple accelerating electrode

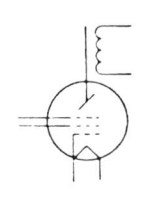

Fuse

general

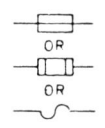

OR

OR

high-voltage fuse

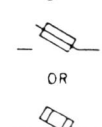

OR

high-voltage fuse, oil

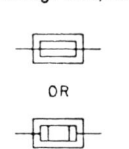

OR

Governor

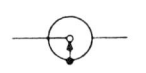

Hall Generator

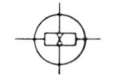

Handset

Hybrid

general

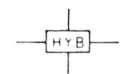

hybrid junction

206

circular hybrid

(Replace asterisk (*) with E, H, or HE to denote transverse field.)

Inductor

general

magnetic-core inductor

tapped inductor

adjustable inductor

continuously adjustable inductor

saturable-core inductor (reactor)

DC WINDING

Key, Telegraph

Lamp

ballast tube

fluorescent lamp, two-terminal

fluorescent lamp, four terminal

cold-cathode glow lamp, ac type

cold-cathode glow lamp, dc type

incandescent lamp

Machine, Rotating

generator

motor

1-phase

3-phase wye grounded

3-phase wye ungrounded

3-phase delta

Magnet, Permanent

PM

Meter

Meter Letter Combinations

(Replace asterisk (*) with proper letter combination.)

A	ammeter
AH	ampere-hour
CMA	contact-making or breaking ammeter
CMC	contact-making or breaking clock
CMV	contact-making or braking voltmeter
CRO	cathode-ray oscilloscope
DB	decibel meter
DBM	decibels referred to one milliwatt
DM	demand meter
DTR	demand-totalizing relay
F	frequency meter
G	galvanometer
GD	ground detector
I	indicating
INT	integrating
μA	microammeter
MA	millammeter
NM	noise meter
OHM	ohmmeter
OP	oil pressure
OSCG	oscillograph, string
PH	phasemeter
PI	position indicator
PF	power factor
RD	recording demand meter
REC	recording

207

RF	reactive factor			
SY	synchroscope			
T	temperature			
THC	thermal converter			
TLM	telemeter			
TT	total time			
V	voltmeter			
VA	volt-ammeter			
VAR	varmeter			
VARH	varhour meter			
VI	volume indicating			
VU	standard volume indicating			
W	wattmeter			
WH	watthour meter			

Oscillator

PAD unidirectional isolator

Path, transmission
general

wire

two conductors

air or space path

dielectric path other than air

crossing of conductors not connected

junction

junction of connected paths, conductors or wires

OR

or only if required by space limitation

shielded single-conductor cable

coaxial cable

two-conductor cable

shielded two-conductor cable with shield grounded

grouping of leads

OR

OR

alternate or conditional wiring

Microphone

Mode Supression

Mode Transducer

Motion Mechanical
translation, one direction

translation, both directions

rotation, one direction

rotation, both directions

Network

NET

208

associated or future wiring

- - - - -

associated or future equipment
(amplifier shown)

circular waveguide

rectangular waveguide

Phase Shifter
general

adjustable

Pickup Head
general

recording

playback

erasing

writing, reading and erasing

stereo

Piezoelectric Crystal

Polarity
positive

+

negative

−

Receiver, Telephone
general

headset

Rectifier
(Represents any method of rectification such as electron tube, solid-state device, electrochemical device, etc.)

general

controlled

bridge type

Relay
alternating current or ringing

fast-operate

fast-release

magnetically polarized

slow-operate

slow-release

Relay Letter Combinations

(Not required with specific symbol.)

AC	alternating current
D	differential
DB	double biased
DP	dashpot
EP	electrically polarized
FO	fast operate
FR	fast release
MG	marginal
NB	no bias
NR	nonreactive
P	magnetically polarized
SA	slow operate and slow release

SO	slow operate
SR	slow release
SW	sandwich wound

Resistor

general

tapped resistor

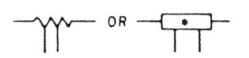

tapped resistor with adjustable contact

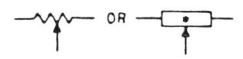

adjustable or continuously adjustable

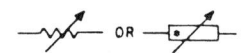

instrument or relay shunt

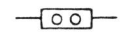

nonlinear resistor

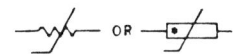

symmetrical varistor

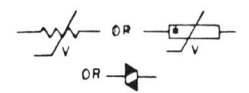

(Replace asterisks (*) with identification of symbol.)

Resonator, Tuned Cavity

Rotary Joint

general (Replace asterisk (*) with transmission-path recognition symbol.)

coaxial in rectangular waveguide

circular in rectangle waveguide

Semiconductor Devices

semiconductor region with one ohmic connection

semiconductor region with plurality of ohmic connections

rectifying junction, P on N region

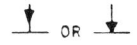

rectifying junction, N on P region

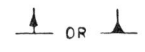

emitter, P on N region

plurality of P emitters on N region

emitter, N on P region

plurality of N emitters on P region

collector

plurality of collectors

transition between regions of dissimilar conductivity

intrinsic region between regions of dissimilar conductivity

intrinsic region between regions of similar conductivity

intrinsic region between collector and region of dissimilar conductivity

intrinsic region between collector and region of similar conductivity

light dependence

temperature dependence

t°

capacitive device

tunneling device

breakdown device

PNP transistor (actual device and construction of symbol)

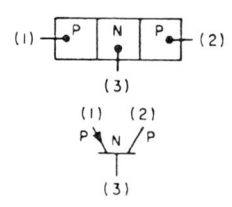

PNINIP device (actual device and construction of symbol)

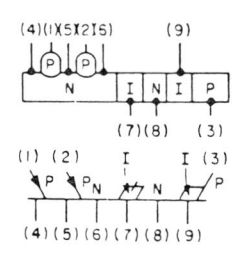

semiconductor diode (also: rectifier)

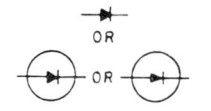

capacitive diode (also: varicap, varactor, reactance diode, parametric diode)

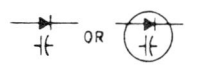

breakdown diode, unidirectional
(also: backward diode, avalance diode, voltage regulator diode, zener diode, voltage reference diode)

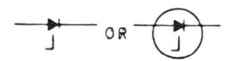

breakdown diode, bidirectional and backward diode (also: bipolar voltage limiter)

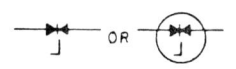

tunnel diode (also esaki diode)

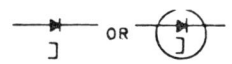

temperature dependent diode

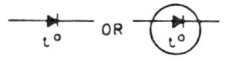

photodiode (also: solar cell)

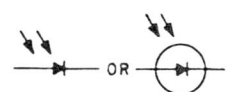

semiconductor diode, PNPN switch (also: Shockley diode, four layer diode)

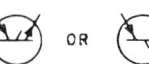

PNP transistor (also: junction, point-contact, mesa, epitaxial, planer, surface-barrier)

PNP transistor with one electrode connected to envelope

NPN transistor (see other names under PNP transistor)

unijunction transistor, N-type base (also: double-base diode, filamentary transistor)

unijunction transistor, P-type base (see other names above)

field-effect transistor N-type base

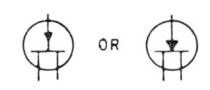

field-effect transistor, P-type base

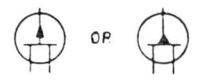

semiconductor triode, PNPN switch (also: controlled rectifier

semiconductor triode, NPNP switch (also: controlled rectifier)

NPN transistor with transverse-biased base

PNP transistor with ohmic connection to intrinsic region

NPN transistor with ohmic connection to intrinsic region

PNN transistor with ohmic connection to intrinsic region

NPIP transistor with ohmic connection to intrinsic region

Shield

Squib
explosive

igniter

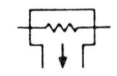

sensing link

Switch
single-throw

double-throw

double-pole, double-throw with terminals shown

with horn gap

knife switch

push button, circuit closing (make)

push button, circuit opening (break)

nonlocking: momentary or spring return—circuit closing (make)

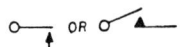

nonlocking: momentary or spring return—circuit opening (break)

nonlocking: momentary or spring return-transfer

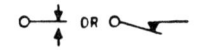

locking—circuit closing (make)

locking—circuit opening (break)

locking—transfer, three position

selector switch

selector, shorting during contact transfer

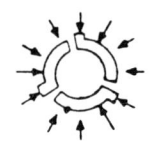

wafer (example shown: 3-pole, 3-circuit with 2 nonshorting and 1 shorting moving contacts)

safety interlock—circuit opening

safety interlock—circuit closing

Switching Function

conducting, closed contact (break)

nonconducting, open contact (make)

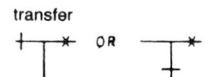

transfer

Synchro
general

Synchro Letter Combinations

CDX	control-differential transmitter
CT	control transformer
CX	control transmitter
TDR	torque-differential receiver
TDX	torque-differential transmitter
TR	torque receiver
TX	torque transmitter
RS	resolver
B	outer winding rotable in bearings

Termination
cable

open circuit

short circuit

movable short

terminating series capacitor, path open

terminating series capacitor, path shorted

terminating series inductor, path open

terminating series inductor, path shorted

terminating resistor

series resistor, path open

series resistor, path shorted

Thermal Element
actuating device

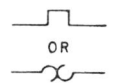

OR

thermal cutout

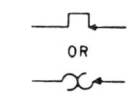

OR

thermal relay

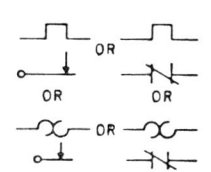

OR

OR

thermostat (operates on rising temperature), with break contact

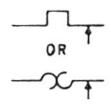

OR

thermostat with make contact

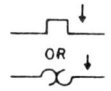

OR

thermostat with integral heater and transfer contacts

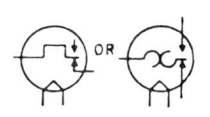

OR

Thermistor
general

with integral heater

213

Thermocouple
general

with integral heater
internally connected

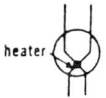

with integral insulated
heater

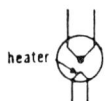

semiconductor
thermocouple, temperature
measuring

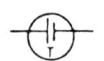

semiconductor
thermocouple, current
measuring

Transformer
general

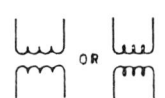

transformer with polarity
marks (instantaneous
current in to instantaneous
current out)

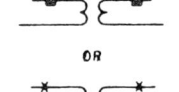

one winding with
adjustable inductance

each winding with
adjustable inductance

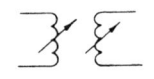

adjustable mutual inductor

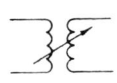

adjustable transformer

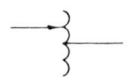

current transformer with
polarity marking

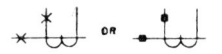

bushing-type current
transformer

potential transformer

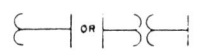

**Transformer
Connection
Winding**
3-phase 3-wire delta or
mesh

3-phase 3-wire delta
grounded

3-phase open delta
grounded at common
point

3-phase wye or star
ungrounded

**Terminal Board
or Strip**

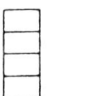

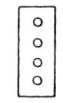

Vibrator
shunt drive

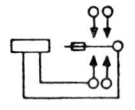

separate drive

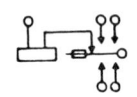

**Visual Signaling
Device**
annunciator, general

annunciator drop or
signal, shutter type

214

annunciator drop or signal, ball type

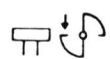

manually restored drop

electrically restored drop

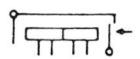

switchboard-type lamp

indicating lamp

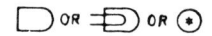

jeweled signal light

Indicating Light Letter Combinations

(Replace asterisk (*) with proper letter combination.)

A	amber
B	blue
C	clear
G	green
NE	neon
O	orange
OP	opalescent
P	purple
R	red
W	white
Y	yellow

Logic Symbols

(Including some duplicate general-purpose symbols; do not mix left-hand symbols and right-hand symbols.)

AND

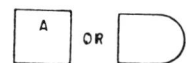

OR

Exclusive or

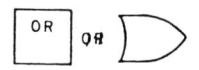

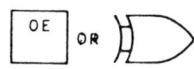

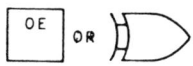

Logic Negation

Electrical Inverter

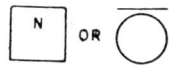

Flip-Flop Complementary

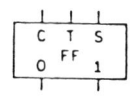

Flip-Flop Latch

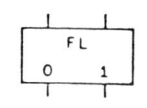

Single Shot

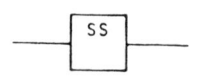

Schmitt Trigger

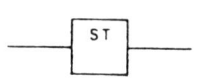

Amplifier

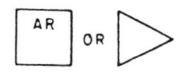

Time Delay

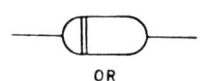

Oscillator

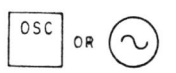

Logic Functions Not Otherwise Symbolized

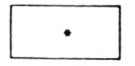

(Replace asterisk (*) with abbreviation or complete identification.)

Application of Relative Level Symbols

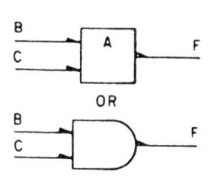

(AND symbol with input 1-states at the more positive level and output 1-state at the less positive potential.)

215

Glossary of Electrical Terms

adjustable resistor—A resistor whose resistance can be changed mechanically. Also adjustable voltage divider.

air capacitor—A capacitor whose dielectric is air.

air core—Descriptive term for coils or transformers that have no iron in their magnetic circuits and are used chiefly in alternating current circuits.

American wire gauge(AWG)—The standard system used for measuring wire size by diameter. The system is also referred to as the Brown & Sharpe (B & S) wire gauge. Wire sizes range from No. 0000, which is the largest, 0.46 in. in diameter, to No. 40, which is 0.003 in. in diameter.

ammeter—An instrument designed to measure current flow in amperes in an electrical circuit. Ammeters are available for either alternating or direct currents. A milliammeter measures current flow in milliamperes and a microammeter measures current flow in microamperes.

ampere—A practical unit of current. It is the current flowing through 1 ohm of resistance at a potential of 1 volt. The movement of 6,280,000,000,000,000,000 electrons at a uniform rate past a given point in a circuit in one second corresponds to a current of one ampere (amp).

ampere-hour—Unit of electrical charge (3600 coulombs) chiefly used to measure the capacity of storage or dry batteries. Specifically, one ampere of current flowing for one hour equals one ampere-hour charge.

Ampere's law—A law of electromagnetics that states that a wire carrying a current in a magnetic field is pushed sideways across the field. The effect is greatest when the directions of the current and field are at right angles to one another. The force depends upon the current and the strength of the field.

ampere-turn—A unit of magnetomotive force obtained by multiplying the current in amperes by the number of turns in a coil.

atom—The smallest unit of any of the chemical elements. Atoms consist of systems of fundamental particles, such as protons, neutrons, electrons, etc. These are arranged with the protons and neutrons compressed into a tiny nucleus at the center with the electrons grouped around it. Atoms of each element have a characteristic structure. The number of protons is the same as the number of electrons. This is the atomic number.

attenuation—Reduction in the strength of an electrical impulse.

AWG—Abbreviation for American wire gauge.

bathtub capacitor—A capacitor enclosed in a metal can with rounded corners like a bathtub.

battery—Two or more cells connected together to serve as a dc voltage source. Loosely, a single cell is sometimes called a battery.

battery charger—A dc generator, rectifier unit, or any similar electrical device used for recharging a storage battery. Current is caused to flow through the cells in a direction opposite to normal current flow.

bell wire—A common term used for the cotton-covered B & S No. 18 copper wire used for making doorbell and thermostat connections in homes.

bleeder current—A current drawn continuously from a power supply to improve its voltage regulation or to increase the voltage drop across a particular resistor.

bleeder resistor—A resistor used to draw a fixed bleeder current. Also used as a safety measure to discharge filter capacitors after the circuit is deenergized.

block diagram—Simplified outline of an electrical or electronic system in which complete circuits or parts are shown as rectangles.

braided wire—Flexible wire made up of small strands woven together.

branch circuits—Circuits containing junctions of three or more conductors. Examples are circuits with two or more components in parallel, bridges, filters, etc. A branch is the path between any two junctions in the circuit.

B & S gauge—Brown and Sharpe wire gauge, the standard gauge used in the United States to specify wire sizes.

C—The symbol used to designate a fixed or variable capacitor in a schematic diagram.

Cable—One or more insulated or noninsulated wires used to conduct electrical current or impulses. Grouped insulated wires are called a multiconductor cable.

capacitance—The quantity of electrical charge that can be received by a system of insulated conductors from a potential source of given value. The term "capacity" is often used in this connection. A capacitor does not become filled, but will receive more charge with increasing potential until breakdown occurs. Unit of capacitance is the farad.

capacitor—An electrical device consisting of two conducting surfaces separated by an insulating material such as air, oil, paper, glass, mica, or ceramic that is capable of storing electrical energy. The capacitance of a parallel-plate capacitor in air is equal to the area of the dielectric divided by 4π times the thickness of the dielectric.

carbon—An element found in graphite or diamond and obtained artificially as lampblack, charcoal, coke, etc. Used in resistors, potentiometers, microphones, battery electrodes, arc lamps, motor and generator brushes, etc.

carbon resistor—A resistor made of carbon particles and a ceramic binder molded into a cylindrical shape with connecting leads attached to opposite ends.

cells—A single unit capable of serving as a dc voltage source. A primary cell, such as a dry cell, cannot be recharged when exhausted. A secondary cell, such as the cell of a storage battery, can be recharged by passing a current through it in a reverse direction.

ceramic—A clay-like material consisting mainly of aluminum and magnesium oxides, that, after molding and firing, is used as insulation. It will withstand high temperatures and is less fragile than glass. Glazed ceramic is porcelain.

ceramic capacitor—A capacitor with a ceramic dielectric.

charge—Electrostatic charge is the quantity of electricity maintained on an insulated object or in a capacitor. When an object has more electrons than normal, it has a negative charge. When an object has less electrons than normal, it has a positive charge. The term "charge" also refers to energy stored in a battery in the form of chemical energy but available as electrical energy.

circuit—A path over which an electric current can flow.

circuit breaker—An electromagnetic device for opening a circuit if the current magnitude exceeds a predetermined value.

circular mil—A unit of area equal to the area of a circle whose diameter is 0.001 in. or 1 mil. The area in circular mils is the square of the diameter in mils. Used chiefly in specifying cross sections of round conductors.

closed circuit—An electrical circuit through which current can flow (e.g., when a power switch is turned to the "on" position). The opposite of an "open" circuit.

coil—A number of turns of wire wound on an iron core, on a coil form made of insulating material, or so as to be self-supporting. A coil offers opposition to any change in current flow.

coil form—The tubing or solid object on which a coil is wound. It can have any shape and can be made from any insulating material, such as paper, cardboard, fiber, bakelite, plastic, ceramic material, or wood.

collector rings—Slip rings on an ac generator.

conductance—The measure of the ability of a component to conduct electricity. In dc circuits it is simply the reciprocal of resistance. Conductance is expressed in mhos. The symbol is G.

conductivity—The specific conductance of a unit specimen of a material. Reciprocal of resistivity.

conductor—A solid liquid or gas that offers little opposition to the continuous flow of current.

connector—A device for electrically interconnecting one or more cables or electronic circuits. There are two main classifications of connectors: male and female. A female connector has one or more contacts set in one or more recessed openings. The recessed opening or openings accommodate a male connector to make electrical connection. The male connector has one or more contacts extending from the connector housing. A line-cord plug end is a simple male connector and a wall electrical outlet is a simple female connector. Male connectors are often called plugs, and female connectors are often called jacks, sockets, or receptacles.

contact emf—A small voltage established when two conductors of different materials touch. It is due to the difference in work functions, or the ease with which electrons can cross the surface boundary in the two directions. Contact emfs cannot be used as a source of electrical energy except by employing two or more junctions at different temperatures, as in a thermocouple.

core—The center of a coil.

coulomb—Unit of quantity of electricity or charge. One coulomb is equal to a movement of 6.3×10^{18} electrons past a point in one second, or one ampere per second.

coulomb's law—A relation applied to electrostatics and magnetics, which states that the force of attraction or repulsion between two charges (or magnetic poles) is directly proportional to the product of the charges (or pole strengths) and inversely proportional to the square of the distance between them, and is modified by the dielectric (permeability) of the material between them.

counter emf—A flow of current in an inductor building up a voltage that tends to push in the direction opposite to the impressed voltage and opposing the original current flow. Also known as back emf.

current—The movement of electrons through a conductor. Current is measured in amperes, milliamperes, and microamperes.

dc—Abbreviation for direct current.

dcc wire—Abbreviation for double cotton-covered wire.

dc generator—A rotating machine that converts mechanical energy into direct-current electrical energy by using the principle of magnetism.

dc resistance—The resistance (opposition to current flow) offered by a circuit or component to dc current flow. Usually called resistance and measured in ohms.

detent—A mechanism used on indexed rotary switches to hold switch firmly in each position. Generally consists of a spring-loaded ball that falls into place in notches or indentations on a plate that rotates with the switch shaft.

dielectric—The insulating material between the plates of a capacitor, between adjacent wires in a cable, or between any two parts of an electronic circuit; it is generally air, mica, paper, oil, cloth, ceramic, or glass.

dielectric constant—The relative permittivity of the dielectric material compared to a vacuum. It is measured by determining how many times greater the capacitance of a capacitor is with the dielectric between the plates than with air. Transformer oil has a dielectric constant of about 2, and mica has a dielectric constant of 5 to 6.

dielectric loss—Energy loss in the dielectric of a capacitor due to hysteresis effects. The losses show up as heat. The effects are analogous to magnetic losses in steels. Heating of the dielectric increases the power factor, which in turn increases the heat losses.

dielectric strength—The maximum voltage that a dielectric can withstand without rupture. Also called insulating strength. Expressed in V solidus mm. The dielectric strength of air is 4000, oil 16,000, and mica 50,000.

direct current—An electric current that flows in only one direction.

disc capacitor—A small, thin, disc-shaped capacitor with radial leads.

discharge—1. Releasing the energy stored in a capacitor by shorting the capacitor terminals. Also glows, arcs, sparks, or corona that occur through gases or vapors at, below, or above atmospheric pressure. 2. Converting the chemical energy of a battery into electrical energy by allowing the battery to send current through a circuit.

dissipation—Unusable or lost energy, such as the production of unused heat in a circuit.

distributed inductance—The inductance that exists along the length of a conductor, as distinguished from inductance concentrated or lumped in a coil.

dpdt—double-pole, double-throw (as used in a switch or relay).

dpst—Double-pole, single-throw (as used in a switch or relay).

dropping resistor—A resistor used to decrease the voltage in a circuit or from which a potential difference (voltage) is taken.

dry cell—A type of primary cell in which the electrolyte is in the form of a paste rather than a liquid. Dry cells are used extensively in radio and electronics batteries.

dry electrolytic capacitor—An electrolytic capacitor in which the electrolyte is a paste rather than a liquid.

E—Symbol for voltage.

Edison storage cell—A storage cell having iron-oxide negative plates and nickel-oxide positive plates immersed in an alkaline solution.

efficiency—the ratio of energy output to energy input, usually expressed in percent. A perfect electrical device would have an efficiency of 100%.

electrical center—The point approximately midway between the ends of an inductor or resistor that divides it into two equal electrical values, as to voltage, resistance, inductance, or number of turns.

electrical inertia—Inductance: opposing any change in current flow.

electric field—A region in space in which a charge experiences a force. Example: a region surrounding a charged object. Field strength or field intensity is defined as the force per unit charge placed in the region as a test charge. The direction of the field is that direction in which a free positive charge would move. Lines drawn in these directions are called lines of force. The density of these lines indicates at each point the strength of the field. A varying electric field is associated with a varying magnetic field, as with radio waves or accelerated electrons. Uniformly moving electrons produce a constant magnetic field, as with electrons flowing in a conductor.

electricity—The flow of electrons in or on a conductor or through a gas or space, as in a vacuum tube.

electrode—1. One of the essential components inside a vacuum tube, transistor, or microcircuit. 2. One of the plates of dry or wet batteries. 3. One of the plates of an electrolytic capacitor. 4. One of the carbons of an arc lamp or furnace.

electrode capacitance—The capacitance between one electrode and all the other electrodes connected together.

electrode potential—The voltage appearing at an electrode.

electrolyte—The liquid, chemical paste, or other conducting medium used between the electrodes of a dry cell, storage cell, or electrolytic capacitor.

electrolytic cell—A cell consisting of a conducting liquid (electrolyte) and two identical electrodes. Such a cell cannot serve as a source of electrical energy, but it can conduct current from an outside source (electrolytic action). Used in electroplating, electroforming, production of gases, and accomplishment of many industrial processes, such as the refinement of metals. If, as a result of current flow, the electrodes become dissimilar, voltaic action becomes possible.

electrolytic capacitor—A fixed capacitor in which the dielectric is a thin film of gas formed on the surface of an aluminum electrode by a liquid or paste electrolyte.

electromagnet—A coil of wire usually wound on an iron core that produces a strong magnetic field when current is sent through the coil.

electromagnetic field—A magnetic field located at right angles both to the lines of force and to their direction of motion.

electromagnetism—Magnetic effects produced by currents rather than by permanent magnets.

electromotive force—Voltage—the force which causes current to flow in a circuit. Particularly, the no-load voltage of a generator or battery.

electron—The elementary charge of negative electricity. Its mass is 9.1×10^{-28} grams. Its charge is -1.6×10^{-19} coulombs. Its diameter is bout 10^{-13} cm.

electronics—A broad term used to cover a field dealing with the use, characteristics and properties of electrons, especially in vacuum radar, control circuits, etc.

electronvolt—Unit of energy. The energy possessed by an electron that has fallen through a difference of potential of one volt.

electrostatic—Pertaining to electricity at rest.

electrostatic charge—An electric charge stored in a capacitor or on the surface of an insulated object.

electrostatic field—The region near an electrically charged object. See **electric field.**

element—1. In chemistry one of the more than 100 primary substances that cannot be divided by chemical means into simpler substances. 2. Any component part of a vacuum tube. 3. One of the conductors (other than the lead-in) of an antenna.

emf—Abbreviation for electromotive force or voltage.

enameled wire—Wire coated with an insulating layer of baked enamel.

energy—Capacity for performing work. Energy in motion is called kinetic energy. Positional energy is called potential energy.

energy levels—Energy states in an atom in which electrons can exist within the atom. Only a certain number of electrons can be in each energy level. Electrons can be raised to higher levels by receiving precisely the necessary energy. These same precise amounts of energy appear as radiation when an electron fails to a lower level. Such energy transfers are the basis of gaseous discharges. The ionizing potential is a measure of the energy needed to remove an electron from an atom.

equivalent circuit—An electrical circuit used to analyze the function of electronic devices such as transistors and microcircuits.

The circuit consists of series and parallel capacitors, inductors, and resistors that are equivalent to the internal capacitance, inductance, and resistance of the device under study.

farad—Unit of capacitance. In the practical system of units, the farad is too large for ordinary use, and capacity measurements are made in terms of microfarads and picofarads. The microfarad is one millionth of a farad, and the picofarad is one millionth of a microfarad. Both are commonly used in radio and electronics.

faraday—A unit of electricity equal to 96,500 coulombs. This is the quantity of electricity that will alter one gram-equivalent weight of matter at each electrode during electrolysis.

field—The effect produced in surrounding space by an electrically charged object.

fixed capacitor—A capacitor with a definite value that cannot be changed.

fixed resistor—A resistor with a definite value that cannot be changed.

flux—1. A material used to promote fusion or joining of metals in soldering, welding, or smelting. Rosin is widely used as a flux in electrical and electronic soldering. 2. A general term used to designate collectively all the electric or magnetic lines of force in a region.

flux density—The number of electric or magnetic lines of force cutting the unit area at right angles.

flux linkage—Magnetic lines of force linking a coil of wire.

fuse—A protective device consisting of a short piece of wire that melts and breaks when the current through it exceeds the rated value of the fuse.

fuse block—An insulating base on which are mounted fuse clips or other contacts for holding fuses.

fuse clip—A spring contact for holding a cartridge fuse in position and providing connections.

fuse wire—Wire made from an alloy that melts at a relatively low temperature.

galvanometer—A D'Arsonval laboratory instrument usually of suspension type for measuring or indicating extremely small electric currents. Its scale usually indicates relative deflection and the actual current, voltage, or charge value must be calculated.

generator—A rotating machine that converts mechanical energy into electrical energy. Also, a radio device or circuit that develops an ac voltage at a desired frequency.

gilbert—Electromagnetic unit of magnetic potential. Expressed in ergs per unit pole. It is the unit of magnetomotive force. Related to ampere turns as follows: 1 gilberts equals $0.4 \times \pi$ ampere turn.

graph—A pictorial representation of the relation between two or more variable quantities.

graphite—One of the crystalline forms of carbon. It is a fair conductor and is used in the construction of carbon resistance elements. Also used in powder form or mixed with grease as a lubricant.

h, hy, or H—Abbreviation for henry, the unit of inductance.

half-cell—An electrode submerged in a suitable electrolyte and used for measuring single-electrode potentials.

henry—The practical unit of self-inductance or mutual inductance. The inductance in which a current changing its rate of flow by one ampere per second induces an electromotive force of 1 volt. Abbreviated h, hy, or H.

hookup—A diagram showing circuit connections for radio or electrical equipment.

hookup wire—Usually tinned and insulated No. 18 or 20 soft-drawn copper wire. Used in wiring radios, TV sets, switchboards, etc. May be solid or stranded.

induced—Produced as a result of exposure to the influence or varation of an electric or magnetic field.

induced charge—An electrostatic charge produced on an object by an electric field in the vicinity.

induced voltage—A voltage produced in a circuit by changes in the number of magnetic lines of force linking or cutting across the conductors of the circuit.

inductance—That property of a coil or other part that tends to prevent any change in current flow. Inductance is effective only when varying or alternating currents are present.

induction—The process by which an object is electrified, magnetized, or given an induced voltage by exposure to a magnetic field.

inductor—A circuit component designed so that inductance is its most important property. Also called a coil.

instrument shunt—A special type of low-value resistor connected in parallel with a meter to extend its current range.

insulated carbon resistor—A carbon resistor encased in a molded ceramic fiber, plastic, or other insulating material.

insulated wire—A conductor of electricity covered with a nonconducting material.

insulating strength—A measure of the ability of an insulating material to withstand electric stress without breakdown.

insulation—Any nonconductive material used to prevent the leakage of electricity from a conductor. Some common insulating materials are cotton, silk, rubber, glass mica, and ceramics.

insulating resistance—The electrical resistance between two conductors separated by an insulating material.

insulator—A device with a high electrical resistance, used for supporting or separating conductors to prevent undesired flow of current between conductors or other objects.

interlock—A safety device that automatically opens the supply current when the cover or door providing access to the equipment is opened.

internal resistance—The resistance of a battery, generator, or circuit component. It acts as a series resistance.

ion—An atom or molecule with fewer or more electrons than normal. A positive ion is one that has lost electrons; a negative ion is one that has acquired electrons.

ionization—The breaking up of a gas atom into a free electron and a positively charged ion. This process corresponds to the "breakdown" of an insulator. Ionization makes a gaseous tube more conductive than an equivalent vacuum tube.

IR drop—The voltage drop produced across a resistance R by the flow of current I through the resistor.

I^2R—Power in watts expressed in terms of current I and resistance R.

I^2R loss—Power loss in transformers, generators, connecting wires, and other parts of a circuit due to current flow I through resistance R of copper conductors. Also called copper loss.

JAN specification—Joint Army-Navy specification.

joule (J)—A unit of energy or work. The absolute joule is equal to 10 million ergs. The international joule is equal to the work required to maintain a current of one ampere for one second in a resistance of one ohm. One joule is equal to one watt-second.

jumper—A short length of conductor used to make a temporary electrical connection.

junction—A point in a circuit where two or more wires are connected.

keeper—An iron or steel bar placed across the poles of a horseshoe magnet to prevent gradual demagnetization by providing a low-reluctance path for the magnetic circuit.
kilo—Metric prefix meaning 1000.
kilohm (kΩ)—Unit of resistance equal to 1000 ohms.
kilovolt (kV)—Unit of voltage equal to 1000 volts.
kilovolt-ampere (kVA)—1000 volt-amperes.
kilowatt (kW)—A unit of electrical power equal to 1000 watts.
Kirchhoff's current law—A fundamental electrical law stating that the sum of all currents flowing to a point in a circuit must be equal to the sum of all currents flowing away from that point.
Kirchhoff's voltage law—A fundamental electrical law stating that the sum of all voltage rises in a complete circuit must equal the sum of all voltage drops in that same circuit.
knife switch—A switch with one or more flat metal blades, each pivoted at one end, as the moving parts. The blades are usually copper. When the switch is closed, they make contact with flat, gripping spring clips and complete the circuit.
kV—Kilovolt.
kVA—Kilovolt-ampere.
kW—Kilowatt.
kWH—Kilowatt-hour.

L—Symbol for inductance.
law of electric charges—Like charges repel, unlike charges attract.
law of magnetism—Like poles repel, unlike poles attract.
layer winding—A coil-winding method in which adjacent turns are laid evenly side by side along the length of the coil form.
lead-acid cell—The most common type of storage or rechargeable cell. The electrolyte is sulfuric acid, and plates are spongy lead. Storage batteries consist of two or more of these cells.
left-hand rule—A rule for determining direction of magnetic lines of force around a single "hot" current-carrying wire. If the fingers of the left hand are placed around the wire that the thumb points in the direction of electron flow, the fingers will point in the direction of the magnetic field.
load—1. The power delivered by a device. 2. A device placed in a circuit to absorb or convert power.
loss—Energy dissipated without accomplishing useful work.
lumped inductance—Inductance concentrated in a component, as distinguished from stray or distributed inductance.

mA—Abbreviation for milliampere.

magnet—A body that attracts iron and steel and, if free to move, sets itself in a definite direction due to the influence of the earth's magnetic field.

magnetic circuit—A closed path of magnetic flux. See **Ohm's Law for magnetic circuits**.

magnetic field—A region in space surrounding a magnet or a conductor through which current is flowing.

magnetic flux—The sum of all the magnetic lines of force from a magnetic source.

magnetic leakage—Passage of magnetic flux outside of the path along which it can do useful work.

magnetic lines of force—Imaginary lines used for convenience to designate the directions in which magnetic forces are acting throughout the magnetic field associated with a permanent magnet, electromagnet, or current-carrying conductor.

magnetic poles—Regions in a magnet near which the field is concentrated, usually the two ends of a magnet (north and south poles). The pole strengths are equal.

magnetic pole strength—Force exerted on a metallic object by a magnetic pole, measured in unit poles. A unit pole is one that repels a similar pole at a distance of one centimeter with a force of one dyne.

magnetics—The branch of science dealing with the laws of magnetic phenomena.

magnetic saturation—That condition in an iron core in which further increases in magnetizing force produce little or no increase in magnetic flux density.

magnetic shield—A soft iron housing used around delicate instruments or radio parts to protect them from effects of stray magnetic fields.

magnetite—An oxide of iron, strongly attracted by a magnet. Magnetite is sometimes polarized and is then called lodestone.

magnet keeper—A bar of soft iron placed across the poles of a horseshoe magnet to complete the magnetic circuit when the magnet is not in use and avoid demagnetizing.

magnetomotive force (mmf)—Magnetic potential difference expressed in gilberts (i.e., ergs per magnet pole). Another unit is the ampere-turn, in which the magnetomotive force is the number of amperes of current in a coil multiplied by the number of turns.

maxwell—The electromagnetic unit of magnetic flux. It is equal to one gauss per square centimeter or to one magnetic line of force.

meg—Sometimes used as an abbreviation for megohm.

meg or mega—A prefix meaning one million.

megger—A high-range ohmmeter with a built-in, hand-driven generator as a direct voltage source, used for measuring insulation resistance values.

mercury—A heavy, silvery colored metal that is liquid at ordinary room temperatures. When heated, it gives off a vapor that is highly conductive when ionized.

mercury battery—A type of battery specially characterized by extremely uniform output voltage throughout its life. Whereas the output voltage of most batteries tends to gradually lessen during their lifetimes, a mercury battery delivers practically constant output voltage until it reaches almost the very end of its life. Also characterized by very long shelf life and high resistance to chemicals, heat, and humidity. Mercury batteries employ a zinc power anode; the cathode is mercury-oxide powder and graphite powder.

mercury switch—An electric switch made by placing a large globule of mercury in a glass tube with electrodes arranged in such a way that tilting the tube causes mercury to make or break (close or open) the circuit.

meter—1. A device that measures and registers the integral of an electric quantity. Example: a watt-hour meter. 2. Any type of electric measuring instrument, such as a voltmeter, ammeter, wattmeter, or ohmmeter. 3. The basic unit of length in the metric system. Equal to 39.37 inches.

mho—The unit of conductance or admittance. It is the word "ohm" spelled backwards.

mica—A transparent flaky mineral that splits readily into thin sheets and has excellent insulating and heat-resisting qualities. It is used extensively to separate the plates of capacitors, to insulate electrode elements of vacuum tubes, and for many other insulating purposes in radio and electronics.

mica capacitor—A type of fixed capacitor with mica as the dielectric.

micro (μ)—A prefix meaning one millionth.

microammeter—A meter reading in microamperes.

microampere—One millionth of an ampere.

microfarad (μF) One millionth of a farad.

microhenry—One millionth of a henry.

microhm—One millionth of an ohm.

micromho—One millionth of a mho.

micron—A unit of length equal to one millionth of a meter or one thousandth of a millimeter.

microswitch—Trade name for a small unit in which a minute motion makes or breaks (opens or closes) a contact. Actuated mechanism is a bent diaphragm that snaps from one position to another rapidly and positively.

microvolt (μV)—One millionth of a volt.

microvoltmeter—A highly sensitive voltmeter that measures difference of potential in microvolts.

mil—A unit of length equal to one thousandth of an inch, used in specifying diameters of round conductors.

milli—A prefix meaning one thousandth.

milliammeter—A meter calibrated in milliamperes.

milliampere (mA)—A unit of current equal to one thousandth of an ampere.

millimicron—A unit of length equal to one thousandth of a micron, and hence equal to one millionth of a millimeter.

milliohm (mΩ)—One thousandth of an ohm.

millivolt (mV)—A unit of voltage equal to one thousandth of a volt.

millivoltmeter—A voltmeter that measures in millivolts.

milliwatt (mW)—A unit of power equal to one thousandth of a watt.

minus sign (–)—Used to indicate subtraction or a negative value. Serves to indicate negative polarity or the negative terminal of a device.

multimeter—A test instrument with different ranges for measuring voltage, current, and resistance.

mutual coupling—When coils are located so that a varying current in one coil induces a voltage in an adjacent coil, the coils are said to be mutually coupled.

mutual inductance—Inductance between two coils. It is the flux linkage in either coil due to one unit of current in the other.

nc—No connection.

negative—A term used to describe a terminal with more electrons than normal. Electrons flow out of the negative terminal of a voltage source.

noninductive capacitor—A capacitor so constructed that it has practically no inductance.

noninductive circuit—A circuit having practically no inductance.
noninductive load—A load having no inductance.
noninductive resistor—A wirewound resistor constructed to have practically no inductance.
noninductive winding—A winding made so that one turn or section cancels the field of the next adjacent turn or section. For example, the wire may be doubled before winding.
nonmagnetic—Not affected by magnetic fields. Some nonmagnetic materials are glass, wood, copper, and paper.
nonmagnetic steel—Stainless steel containing chromium, manganese, or nickel, all of which are practically nonmagnetic at ordinary temperatures.
normally closed—A term applied to a magnetically operated switching device or to its contacts, specifying that the device contacts will conduct power when the motor element or magnet is not energized. Term is also applied to keys and switches.
normally open—Opposite of normally closed in that the device contacts will not conduct power when the motor element or magnet is not energized.
north pole—That pole of a magnet at which magnetic lines of force are considered as leaving the magnet. If the magnet is free to move, its north pole will point north.
nucleus—The central part of an atom consisting of protons and neutrons. It has a positive charge and constitutes practically the entire mass of the atom.

oersted—The unit of magnetic intensity in dyne's per unit magnet pole.
ohm—The practical unit of electrical resistance. It is that resistance in which one volt will maintain a current of one ampere.
ohmic value—Resistance in ohms.
ohmmeter—An instrument for measuring resistance. It consists essentially of a milliammeter in series with a suitable dc voltage and series resistors.
Ohm's Law—A fundamental electrical law expressing the relationship between voltage, current, and resistance in a dc circuit.
Ohm's Law for magnetic circuits—States that a dc magnetic circuit is analogous to a dc electrical circuit in that magnetic flux is established by a magnetomotive force and is limited by reluctance. Reluctances in series and parallel may be evaluated just as resistances are. Further, the reluctance of a component may

be calculated from the length and cross-sectional area if, in the magnetic case, the permeability is used instead of the conductivity or resistivity.

ohms per volt—A sensitivity rating for voltage-measuring instruments, obtained by dividing the resistance of the instrument in ohms at a particular range by the full-scale voltage value at that range. The higher the ohms-per-volt rating, the more sensitive is the meter.

omega (Ω)—Greek letter representing the word "ohm".

open circuit—A circuit that is not continuous.

open-circuit voltage—The voltage at the terminals of a battery or other voltage source when no current is flowing—when no load is connected across the voltage source.

open core—An iron core fitting inside a coil but having no external return path so that the magnetic circuit has a long path through air.

parallel connection—A shunt connection. A method of connecting two or more components into a circuit side by side so that one line terminal is connected to one end at each component (these ends are thus connected together), and the other line terminal is connected to the other side at each component.

permeability (μ)—The measure of the ease with which a given material can carry magnetic lines of force. Expressed as a multiple of the permeability of air which is 1.

permittivity—Specific inductive capacity. It is a measure of the ease with which a given material can carry electric lines of force. Permittivity of a medium is the reciprocal of the force with which two like unit charges repel one another at a distance of one unit when in the medium. The value depends on the system of units used. Relative permittivity or dielectric constant is the ratio of the permittivity of the material to that of a vacuum.

phenolic material—A thermosetting insulating plastic material available in many different types and sizes.

photocell—See **photovoltaic cell**. Sometimes incorrectly referred to as a phototube.

photoconductive—A substance that changes its electrical conductivity under varying degrees of illumination is said to be photoconductive. For example, selenium has approximately eight times as much resistance in the dark as in the light.

picofarad (pF)—One millionth of a microfarad.

piezoelectric—Pressure electricity. Property of some crystals to generate a voltage when mechanical force is applied, or to produce a mechanical force by expanding or contracting when a voltage is applied.

plastics—A class consisting of a large variety of substances formed by polymerization. Liquids or vapors made up of small molecules are caused to combine chemically into enormous molecules forming highly inert solids with unique insulating properties. Plastics are molded during formation but may be machined later. Many may be softened by heat. Plastics have many uses in electricity and electronics.

polarity—1. An electrical condition determining the direction in which current tends to flow. Applied to dc sources, also to components when connected in dc circuits. 2. The quality of having one positive charge and one negative charge. 3. The quality of having one north magnetic pole and one south magnetic pole.

positive—A term used to describe a terminal with fewer electrons than normal, so that it attracts electrons in seeking to return to its normal state. Thus, electrons flow into the positive terminal of a voltage source.

potential—Voltage. The work per unit charge required to bring any charge to the point at which the potential exists. The number of volts is the joule per coulomb.

potential difference—The difference in voltage at two points.

potentiometer—A popular term for a variable resistor.

powdered iron core—A core consisting of powdered magnetic material mixed in a suitable matrix and pressed into the required shape.

power—Rate of doing work. Energy per unit time. May be expressed in horsepower (550 foot-pounds per second). In electrical field is expressed in watts.

power output—The power in watts delivered by an amplifier to a load.

power rating—The power available at the output terminals of a device such as a tube, transistor, or microcircuit when the device is operated according to the manufacturer's specifications both as to input and output loads.

power ratio—The ratio of the power output to the power input of a device. Usually expressed in decibel loss or gain.

primary battery—A battery of one or more primary cells.

primary cell—A type of cell in which the generated voltage is due to the permanent chemical changes in the cell material. A primary cell cannot be recharged. This is the earliest known type of cell.

proton—A particle with a charge of +1 and a mass of 1 in atomic units. This charge is equal to the magnitude of the charge of an electron. The mass of the proton is 1850 times that of an electron.

RC circuit—Designation for any resistor-capacitor circuit.

RC constant—The time constant of a resistor-capacitor circuit, equal in seconds to the value of the resistance multiplied by the value of the capacitance.

repulsion—A force tending to separate objects or particles with like electrical charges or magnetic polarities.

resistance—The nonreactive opposition that a device or material offers to the flow of current. The opposition results in production of heat in the material carrying the current. Resistance is measured in ohms and is usually designated by the letter R.

resistance drop—Voltage drop occurring between two points due to the flow of current through a resistance connected between those points. Known as IR drop.

resistance loss—Power loss due to current flowing through a resistance.

resistance wire—Wire made from an alloy having high resistivity.

resistivity—The specific resistance of a unit specimen of material. Expressed either as ohm-centimeters or ohms per circular mil foot. The resistive property of a material is thus expressed as the number of ohms in a piece one square centimeter in cross section and one centimeter long, or a piece one circular mil in cross section and one foot long. Examples: The resistivity of copper is 1.72 microhm centimeters or 10.4 ohms per circular mil foot.

resistor—A component that offers resistance to the flow of electric current. Its electrical size is specified in ohms or megohms. A resistor also has a power-handling rating in watts indicating the amount of power that can safely be dissipated as heat by the resistor.

retentivity—A measure of the ability of an iron-alloy material to hold its magnetism.

RETMA—Abbreviation for Radio Electronics Television Manufacturers Association. Now changed to EIA (Electronics Industries Association).

RETMA color code—One of the systems of color markings developed by RETMA for identifying electrical values and terminal connections for electrical/electronic parts.

right-hand rule—A rule for determining direction of magnetic lines of force.

rosin-core solder—Solder made up in tubular form with the inner space containing rosin flux for effective soldering.

rotary switch—A multiposition switch operated by rotating a control knob attached to its shaft.

scc wire—Abbreviation for single-cotton-covered wire.

sce wire—Abbreviation for single-cotton-covering over enamel insulation on a wire.

schematic diagram—A diagram showing electrical connections of a circuit by symbols representing the components.

secondary cell—A dc voltage source capable of storing electrical energy. When exhausted it can be recharged by sending a direct current through it in the reverse direction. Each cell of an ordinary storage battery is a secondary cell.

series—A way of arranging parts in a circuit by connecting them end to end to provide a single path for current flow.

short circuit—A low-resistance connection across a voltage source or between the sides of a circuit or line usually accidental and usually resulting in excessive current flow that often causes damage.

silk-covered wire—Wire covered with one or more layers of fine floss silk.

single-pole switch—A switch that can be closed in only one position, thus always closing the same contact or set of contacts. May be single, double, triple, or multiple pole.

solder—An alloy of lead and tin that melts at a fairly low temperature (about 500°) and is used for making permanent electrical connections between parts and wires.

soldering gun—A device for applying heat to a joint to be made permanent by soldering.

solid conductor—A single wire. A conductor that is not divided into strands.

source—A term sometimes used to describe the path supplying electrical energy to a circuit.

south pole—The pole of a magnet at which magnetic lines of force are assumed to enter; they emerge from the north pole. If the magnet is free to move, its south pole will point to the earth's

north pole. Also called a north-seeking pole.
sphaghetti—Heavily varnished cloth tubing sometimes used to provide insulation for electrical circuit wiring.
spark—A momentary flash due to an electric discharge through air or some other dielectric material.
spdt—Abbreviation for single-pole, double-throw. Applies to a switch or relay contact arrangement.
splice—A connection of two or more conductors or cables to provide good mechanical strength as well as good conductivity.
square mil—An area equivalent to a square having sides 1 mil (0.001 inch) long.
static charge—An electric charge accumulated on an object, usually by friction.
sun battery—A semiconductor device developed by Bell Labs made up of a thin wafer cut from a single crystal of n-type silicon, diffused with a small amount of boron to produce a thin layer of p-type silicon. The resulting cell when illuminated produces electricity. Direct sunlight conversion efficiency = 11%.
switch—A mechanical device for completing, interrupting, or changing the connections of an electrical circuit.
symbol—A simple design used to represent a part in a schematic circuit diagram. A letter used in formulas to represent a particular quantity.

terminal—A fitting for making electrical connections.
test prod—A metal point attached to an insulating handle and connected to a flexible test lead.
thermocouple—A pair of dissimilar metals in contact forming a thermojunction at which a voltage is developed when the junction is heated.
tubular capacitor—A paper or electrolytic capacitor having the form of a cylinder with leads projecting axially from one or both ends.

V—Abbreviations for volt and voltmeter.
variable resistance—A resistor with a sliding contact so that its resistance can be changed.
volt—The practical unit of voltage, potential, or electromotive force. One volt is the electromotive force to send one ampere through a resistance of one ohm.
voltage—The electrical pressure that makes current flow through a conductor.

voltage drop—The voltage developed between the terminals of a circuit component by the flow of current through the resistive element of that part.

voltage rating—The maximum sustained voltage that can safely be applied to or taken from an electrical or electronic device without risking the possibility of breakdown.

voltaic cell—A cell consisting of a conducting liquid and two dissimilar electrodes. Such a cell can function as a source of electrical energy.

voltmeter—An instrument for measuring voltage.

volt-ohm-milliammeter (vom)—A test instrument having provisions for measuring voltage, resistance, and current. It consists of a single meter with the necessary number of scales and a switch that places the meter in the correct circuit for a particular measurement.

W—Designates power in watts.

watt (W)—The practical unit of electric power.

wattmeter—A meter used to measure the power in watts or kilowatts.

winding—One or more turns of wire forming a continuous coil. Also, the coil itself, as in inductors.

wire—A rod or filament forming a metallic conductor with uniform thickness. Used in electricity to provide a path for electric current between two points.

wirewound resistor—A resistor constructed by winding a high-resistance wire on an insulated form. The resulting element may or may not be covered with a ceramic insulating layer.

Index

A
ammeter, 117
ammeter scales, 122
ampere, 22
atoms, 2
attraction
  law of, 10, 63

B
batteries, 83
  parallel connected, 89
  series connected, 87
  series-parallel connected, 90
battery, 20
  dry cell, 21
battery combinations, 87

C
capacitance, 135
  calculating total, 144
  factors affecting, 142
capacitor color code, 150
capacitor dielectric material, 142
capacitor plate spacing, 142
capacitors
  air, 146
  bathtub, 149
  ceramic, 148
  electrolytic, 149
  metal-encased, 149
  mica, 147
  paper, 148
  theory of, 136
  types of, 146
cell
  primary, 83
charging
  constant-current method, 98
  constant-voltage method, 98
chemical action, 91
chemical elements
  table of, 4, 5
chemical energy, 20
circuit
  open, 28
  parallel, 46, 47
circuit analysis, 55
circuits
  electrical, 38
  magnetic, 76, 81
  parallel, 44
  series-parallel, 53
  simple, 26
color code
  capacitor, 150
  resistor, 112, 113
conductor, 8
  magnetic field of a, 77
conservation of energy
  law of, 14
current, 18, 39
current flow, 22

239

D

D'Arsonval meter movement, 117
dc circuit analysis, 31
dc formulas, 185
dc measuring devices, 117
dielectric constant, 142
digital voltmeters, 129
division by powers of ten, 33
dry cell battery, 21

E

efficiency, 15
electrical abbreviations, 191
electrical circuits, 38
electricity
 static, 9
electrolysis, 94
electrolyte, 84, 100
electromagnet, 79
electromagnetism, 76
electromotive force, 18
electronic symbols, 199
electrons, 3
electrostatic field, 11
electrostatic lines of force, 11
emf, 18
 sources of, 18
energy, 12
exponents, 32
 negative, 33
 positive, 33

F

farad, 135
force
 electromotive, 18
 electrostatic lines of, 11
 magnetomotive, 81
formulas
 dc, 185

H

heat, 18
horsepower, 15
hydrometer, 102

I

inductance, 159
 calculating total, 170
 factors affecting, 162
 mutual, 167, 169
induction
 magnetic, 23
inductive time constant, 165
inductor construction, 171
inductors
 parallel-connected, 170
 uses for, 176
insulators, 8
internal resistance, 86
ionization, 85

K

kinetic energy, 14

L

left-hand rule, 77
light, 19

M

magnet
 compound, 72
 horseshoe, 72
 laminated, 72
 ring, 73
magnetic attraction, 60
magnetic circuits, 76, 81
magnetic field, 61, 66
magnetic field of a conductor, 77
magnetic force, 64
magnetic induction, 23
magnetic lines of force, 67
magnetic poles, 61, 64
magnetism, 59
 theory of, 63
magnetomotive force, 81
magnets, 21
 bar, 71
 care of, 75
 shapes and uses of, 70
mathematics, 31
matter, 1
meter movement
 moving coil, 117
meter sensitivity, 117
microfarad, 135
molecules, 2
multimeter, 117, 132
multiplication by powers of ten, 33
mutual inductance, 167, 169

N

negative temperature coefficient, 112
negative terminal, 18

O

Ohm, Georg Simon, 25
 the, 24
Ohm's law, 34, 35, 47, 53, 56
ohmmeter construction, 130
ohmmeters, 117

240

basic multirange, 131
use of, 131
open circuit, 28

P

parallel circuit, 44, 46, 47
  laws for, 49
  simple, 45
parallel resistor combinations, 49
path of least resistance, 46
permeability, 65
photoelectric effect, 19
physical laws, 1
physics, 1
piezoelectric effect, 22
polarity, 17
positive temperature coefficient, 112
positive terminal, 18
power, 14, 44
power memory wheel, 45
primary cell, 83
proton, 3

R

reciprocal method, 49
reluctance, 81
repulsion
  law of, 10, 63
resistance, 24, 25, 39
  internal, 86
resistor color code, 112, 113
resistor tolerances, 110
resistor values, 110
resistors
  carbon-composition, 108
  types of, 107
  variable, 109
  wirewound, 108
retentivity, 59
rule
  left-hand, 77

S

schematic diagram, 28
schematic symbols, 27
secondary cell construction, 94
secondary cells, 90
  care of, 103

ratings for, 97
selenium, 19
self-inductance, 159
sensitivity
  voltmeter, 128
series-parallel circuits, 53
specific gravity, 100
static electricity, 9
symbols
  schematic, 27

T

temperature coefficient, 112
  negative, 112
  positive, 112
thermistors, 112
thermocouple, 19
thermoelectric effect, 19
time constant, 140
  inductive, 165

U

units
  basic measurement, 17

V

varistors, 114
volt, 18
Volta, Allesandro, 18
voltage, 41
voltage drop, 41
voltaic cell, 84, 85
voltaic pile, 84
voltmeter, 126
voltmeter sensitivity, 128
voltmeters, 117
  digital, 129
  loading effect of, 128

W

windings
  flat-spiral, 173
  multilayer, 173
  noninductive, 175
  toroidal coil, 174
wire gauges, 197
work, 12

Other Bestsellers From TAB

☐ **BASIC INTEGRATED CIRCUITS—Marks**

Bored with building the same old digital IC circuits from kits or stock plans available in hobby magazines or ordinary project guides? Then why not expand your horizons and start building your own original IC circuits that you've designed to do exactly what you want them to do! With the step-by-step guidance and over-the-shoulder advice provided by electronics design expert Myles Marks, anyone with basic electronics know-how can design and build almost any type of modern IC device for almost any application imaginable! 432 pp., 319 illus.
**Paper $16.95   Hard $26.95
Book No. 2609**

☐ **THE ENCYCLOPEDIA OF ELECTRONIC CIRCUITS—Graf**

Here's the electronics hobbyist's and technician's dream treasure of analog and digital circuits—nearly 100 circuit categories . . . over 1,200 individual circuits designed for long-lasting applications potential. Adding even more to the value of this resource is the exhaustively thorough index which gives you instant access to exactly the circuits you need each and every time! 788 pp., 1,782 illus. 7" × 10".
Paper $29.95   Book No. 1938

☐ **DESIGNING IC CIRCUITS . . . WITH EXPERIMENTS—Horn**

With this excellent sourcebook as your guide, you'll be able to get started in the designing of your own practical circuits using op amps, 555 timers, voltage regulators, linear ICs, digital ICs, and other commonly available IC devices. It's crammed with practical design and construction tips and hints that are guaranteed to save you time, effort, and frustration no matter what your IC application needs. 364 pp., 397 illus.
**Paper $16.95   Hard $24.95
Book No. 1925**

☐ **ENCYCLOPEDIA OF ELECTRONICS**

Here are more than 3,000 complete articles covering many more thousands of electronics terms and applications. A must-have resource for anyone involved in any area of electronics or communications practice. From basic electronics or communications practice . . . from basic electronics terms to state-of-the-art digital electronics theory and applications . . . from microcomputers and laser technology to amateur radio and satellite TV, you can count on finding the information you need! 1,024 pp., 1,300 illus. 8 1/2" × 11".
Hard $60.00   Book No. 2000

☐ **BASIC ELECTRONIC TEST PROCEDURES—2nd Edition—Gottlieb**

The classic test procedures handbook, revised and expanded to include all the latest digital testing and logic probe devices! It covers the full range of tests and measurements. Clearly spelled out techniques are backed up by explanations of appropriate principles and theories and actual test examples. Plus there are over 200 detailed show-how illustrations and schematic diagrams. 368 pp., 234 illus. 7" × 10".
**Paper $16.95   Hard $23.95
Book No. 1927**

☐ **BASIC ELECTRONICS THEORY—WITH PROJECTS AND EXPERIMENTS—2nd Ed.**

If you're looking for an easy-to-follow introduction to modern electronics . . . or if you're an experienced hobbyist or technician in need of a quick-reference guide . . . there's simply no better sourcebook than this guide. It includes all the basics plus the most recent digital developments and troubleshooting techniques. All new material covers AM, stereo, new video sources, computers, and more. 672 pp., 645 illus.
**Paper $18.95   Hard $29.95
Book No. 1775**

Other Bestsellers from TAB

☐ **THE ILLUSTRATED DICTIONARY OF ELECTRONICS—3rd Edition— Turner & Gibilisco**

The single, most important reference available for electronics hobbyists, students, *and* professionals! Identifies and defines over 27,000 vital electronics terms—more than any other electronics reference published! More than *2,000 new topics* have been added to this state-of-the-art 3rd Edition! *Every* term has been revised and updated to reflect the most current trends, technologies, and usage—with every meaning given for every term! Covers basic electronics, electricity, communications, computers, and emerging technologies! Includes nearly 400 essential drawings, diagrams, tables, and charts! It's the only electronics dictionary that accurately and completely identifies the hundreds of abbreviations and acronyms that have become "standard" in the electronics and computer industries! 608 pp., 395 illus. 7" × 10".
**Paper $21.95   Hard $34.95
Book No. 1866**

☐ **ELECTRONIC COMPONENTS HANDBOOK FOR CIRCUIT DESIGNERS**

Whether you're an electronics novice, a seasoned experimenter or a professional technician, you can find just the data you need in this outstanding new selection. You'll find exact details on what circuit components are, what they do, and how they're used. Plus, you'll get background information on each device and over-the-shoulder advice on how to put them to work. 336 pp., 237 illus.
Paper $13.95   Book No. 1493

☐ **THE COMPLETE BATTERY BOOK—Perez**

Here's the practical, money-saving information you need to choose the right batteries for any job you have in mind *and* show you how to get longer life from each and every battery you use. It even gives you complete schematics for constructing your own low-cost battery chargers and motor generator sets! If you're an electronics hobbyist, this book is packed with information that has immediate practical use in your experiments and projects. If you're a homeowner or RV or boat owner, it's an ideal source for time- and money-saving advice on battery usage. And it's a must for anyone installing an alternative energy system. 192 pp., 111 illus. 7" × 10".
**Paper $16.95   Hard $24.95
Book No. 1757**

☐ **BEGINNER'S GUIDE TO READING SCHEMATICS**

Electronic diagrams can be as easy to read as a roadmap with the help of this outstanding how-to handbook. You'll learn what each symbol stands for and what the cryptic words and numbers with each one mean, as well as the meaning of the lines and other elements and how these are combined into complete circuits. Includes block diagrams to show how sections of complicated circuits relate to each other, pictorials to tell where pieces are physically located, and flowcharts to give you a step-by-step list of what should be happening—both where and when. 140 pp., 123 illus.
Hard $14.95   Book No. 1536

*Prices subject to change without notice.

Look for these and other TAB BOOKS at your local bookstore.

TAB BOOKS Inc.
P.O. Box 40
Blue Ridge Summit, PA 17214

Send for FREE TAB Catalog describing over 900 current titles in print.

DISCARDED

JUN 23 2025